The Rape of the Wetlands

The Rape of the Wetlands

Keith A. Wilkins PhD

Writers Club Press
San Jose New York Lincoln Shanghai

The Rape of the Wetlands

Writers Club Press
an imprint of iUniverse.com, Inc.

For information address:
iUniverse.com, Inc.
620 North 48th Street, Suite 201
Lincoln, NE 68504-3467
www.iuniverse.com

ISBN: 0-595-13952-3

Printed in the United States of America

To my wife, Shirley, with love and thanks.

Epigraph

How can you buy or sell the sky? The land? The idea is strange to us. If we do not own the freshness of the air and the sparkle of the water, how can you buy them? Every part of the earth is sacred to my people. Every shining pine needle, every sandy shore, every mist in the dark woods, every meadow, every humming insect....

If we sell you our land, remember that the air is precious to us. The wind that gave our grandfather his first breath also received his last sigh. The wind also gives our children the spirit of life. So if we sell you our land, you must keep it apart and sacred, a place where man can go to taste the wind that is sweetened by the meadow flowers.

This we know: the earth does not belong to man, man belongs to the earth. All things are connected like the blood that unites us all. Man did not weave the web of life, he is merely a strand in it. Whatever he does to the web, he does to himself.

Chief Seattle's response when President Franklin Pierce stated in 1855 he would buy the lands of the Chief's tribe

Introduction

August 1950——The young boy on his Schwinn slammed on the brakes after turning off the country road. He jumped off, feeling the rough gravel through the soles of his once white sneakers. He pushed his bike along the narrow path that led to the marsh.

Cody Matheson looked forward to talking with the old man and maybe going out in the canoe to explore the marsh's waterway.

But his anticipation turned to concern when he saw the blue smoke rising above the trees and heard the growl of engines. He pushed his bike faster, careful not to hit his leg against the pedal.

He couldn't imagine what was going on. The old man would tell him. He knew all about the marsh. He knew everything.

Cody felt the warmth of the August afternoon sun shining through the trees from a cloudless Massachusetts sky and briefly noted how it laid a carpet of dappled shade along the wooded path. But right now his thoughts were on what was going on ahead.

As before, he had come here to talk with the old man who lived in the shack. He didn't know where the man came from, but that didn't matter. Cody loved listening to him talk about the marsh and the wildlife living in it.

The old man would point out many things, such as the way fiddler crabs scampered over the mud where the cord grass filtered the tides. When Cody paddled the canoe along the meandering channels of the marsh, he felt like an Algonquin surveying his primitive domain; however, a glance toward the town of Marshfield, with its church steeples and gray factories, destroyed the illusion.

"*Most folks don't give a marsh a second look,*" the old man told him. "*There's plenty of life out there. All you got to do is look.*"

After listening to the old man, Cody had come to see the marsh sur-rounded by an aura of both mystery and fascination.

"*Just set quiet, and you'll see,*" the man said. "*They know you're there, all the creatures, the birds, crabs, muskrats, turtles, and the fish. And the mud, oh, the mud's very much alive. 'Course you got to get used to the smell, but after a while it ain't so bad—if you don't mind the odor of rotten egg.*" he mused, half smiling

"*I like it.*"

The old man chuckled. "*And the goddamned mosquitoes.*"

"*Well, I don't like the mosquitoes.*"

The old man's eyes grew distant. "*They're part of it, though. They belong here, like the mud crabs, the cord grass, and all the rest.*" *He paused.* "*What about you? Where do you belong?*"

"*I haven't thought much about that, but I'll never leave. I love the marsh.*"

When Cody observed something different, he told the old man. "*I saw some clam worms the other day, and a mud snail when the tide was out. I saw some sheepshead minnows and sand fiddlers, and there's a marsh wren's nest over there.*"

"*There ain't nothin' like a marsh,*" the man said. "*It makes one hell of a lot of organic matter. "Course, some might say, 'so what?'*"

As Cody hurried, pushing his bike along the path, the roar of engines grew louder and he could smell the acrid odor of something—diesels? His concern grew with each running step.

At the end of the path, he halted abruptly. His deep blue eyes scanned what was left of this part of the marsh. He watched with disbelief as growl-ing bulldozers pushed piles of dirt out into it. Horrified, he saw the wren's nest in the tall cord grass threatened by an encroaching wall of dirt. The birds—herons, plovers, and sandpipers—circled in frightened confusion.

Then he saw what was left of the shack, its boards protruding from a mound mass of dirt pushed up by a bulldozer's blade. Where was his canoe?

Stunned, Cody threw down his bike.

"No!" he screamed, and ran at the massive bulldozer grinding toward the wren's nest. He stopped in front of its towering blade and threw up his arms.

The machine snorted to a halt. The startled driver jumped to his feet. "Get the hell away, kid."

"No, no!" Cody waved his arms.

The other men, wearing yellow safety hats, stared at him.

The driver shouted. "Someone get this damn fool out of here."

Cody looked up at a big man wearing a red construction hat with the word 'foreman' on it.

"What the hell's going on?" the foreman demanded, as he approached.

"This kid came racin' out of them woods," the driver said. "I could have hit him."

The big man scowled down at the boy. "What's the idea?" The others regarded the boy with curiosity.

"You're killing them."

"Killing what?"

"The nest. There's a nest. Right over there." Cody pointed and ran a hand through his blond hair.

"And what have you done with him?"

"Who?"

"The man who lives here."

"They took him yesterday."

"Where did they take him?" He looked about frantically, tears in his eyes. "What are you doing? Why'd you smash the hut, and where's my canoe?"

The foreman jerked his head. "That it over there?" He looked down at the boy and grabbed his arm. "We'll look after it until someone picks it up."

"Leave me alone." Cody wrenched free.

"Damn it, son, I'm taking you home." The foreman stepped toward him and caught him again. "Where do you live?"

The boy struggled and yelled to be released. "You're destroying it. Stop! Please stop." He sobbed in anger.

"You better get on home," one of the men said.

"You can't run over the nest. Don't do this."

"We got work to do," the foreman said. "Besides, what's it to you?"

"Don't you know anything about a marsh?" The boy felt ashamed for crying in front of the men.

"Listen, kid," the foreman told him, "we got our orders. Now tell us where you live."

The big man relaxed his hold. Taking advantage, Cody yanked free and ran for the path. The men watched him, but did not go after him. "You stay away from here," the foreman called after him.

Cody grabbed his bike, but before rushing it up the narrow path, he stopped and turned. "Goddamn, you'll pay for this."

CHAPTER ONE

May 1990—Cody listened to the wind, howling through the rigging like the hysterical voice of the demented storm. He swore, and called himself stupid for not lowering the sails when he'd seen the dark lowering clouds from which lightning forked into the sea.

He grasped the wheel and struggled to keep the boat's heading into the gale. The bow plunged into each trough, sending a huge wave crashing over the boat and drenching him.

Now, in building seas, he had no choice but to climb forward to pull down the sails. He started the engine in order to have headway, and secured the helm, otherwise the boat could swing around broadside to the wind and waves. A prescription for broaching. Now he was able to lower the jib, which could be done from the Columbia's cockpit.

As he released the jib sheets from their cleats, he activated the roller furling. That done, he glanced at the waves washing over the deck. It would be slippery. He reached for the grabrails and moved slowly forward. At the mast he fastened his safety harness around it. He sure in hell didn't want to be thrown into the raging sea and drown. If that happened, he'd never be available to help Mitch with the trouble he was in.

Each wave heaved up the bow only to pass and let the boat plunge into the following trough like a shuddering bull, throwing up a geyser of water that cascaded on the deck. Matheson envisioned the boat plowing to the bottom of the Atlantic. But each time the Columbia shook itself free from the ocean's clutches and climbed the next wave. Cody clung frantically to the gyrating boom. The boat, he knew, could withstand a greater punishment than he.

Before the storm, he had been lulled into a false sense of security. After leaving Norfolk, Virginia, four days ago, telling his lawyer he was not waiting for the divorce to be final, he had enjoyed great sailing weather: a firm breeze and gentle swells.

As Cody held on with one hand, he released the mainsail halyard and let the line play through his hand. He had to release this line so he could haul down on the forward edge of the sail. He swore aloud, as he struggled against the force of the wind putting heavy pressure on the sail hanks.

He thought trying to stay on a barroom bronco would be child's play compared to this.

Finally, exhausted, he got the sail wrapped around the boom and secured the bungee straps to prevent the sail from billowing loose in the wind.

He hugged the boom as he gasped for breath. Then he began to work his way back to the cockpit. First, he released his safety harness and clipped it to the starboard life line. As he inched aft, the Columbia lurched. One of three wire shrouds securing the mast tore loose from its chainplate and flayed in the wind. Cody grabbed for it, but he missed. The turnbuckle struck him, slashing open his forehead. Stunned, he fell. As he did, he grabbed for one of the other shrouds, but missed. He pitched over the starboard side.

"Jesus Christ," he shouted, "this is it."

Although he knew the safety harness would prevent his being lost to the sea, he had, in that frantic effort to save himself, caught hold of the starboard life line. Pain shot through his wrenched shoulder. Despite the agony, he hung on, his body dragging through the water. The Columbia bucked and heaved as though trying to shake him loose.

He dared not let go, because, although he'd be hauled along by the line, he'd have to pull himself hand over hand until he could grab anything to haul himself aboard. If he had the strength left to do so.

"Goddamn," he cried.

That he fleetingly wondered how his ex-wife would react to his drowning, struck him as singularly odd. Why the hell should he wonder how she felt?

His thoughts flashed back, as though he were going through an after-death life review. He thought about his father, who had been foreman in a shoe factory, and his parents' love of the ocean, and the times he and his mother and father went sailing on Cape Cod bay, and how his mother shook her head over the fact that her husband never understood the geometry of sailing and never accepted that sheets were lines, not canvas

He recalled the news that the family sloop had disappeared off Provincetown, with no trace of his parents. Cody snapped back to his present ordeal. What a damn fool to be wasting time and energy thinking of the past. He had to focus on saving himself. He had to get back aboard—somehow

He waited for the deck to dip into the water as the Columbia rolled, then, with a desperate effort, threw one leg up over the side and hooked his knee about a stanchion. He was tiring rapidly.

Slowly, he pulled himself up until he could slide under the lower lifeline and onto the deck. On his hands and knees, he watched his blood splatter on the deck to be washed away in the rain, and wished he could sink into the blackness that hovered on the periphery of his blurred vision.

Lightning stabbed the turbulent waters around him, like the spears of Odin. One hit the top of a near-by wave and sent a pyrotechnic display of sparks scattering across the violent water. Although the tall mast was grounded, Cody knew that it could attract lightning. If the lightning didn't kill him, it could set the boat afire.

At last, struggling to his feet, he worked his way back to the cockpit. Once behind the wheel, he reattached his lifeline and released the steering lock. He then eased the throttle forward to give the boat greater headway through the huge waves, continuing to rush upon him and crash across the deck, sending water flooding the cockpit.

His vision dimmed as the wild seas wrestled the boat. Was he losing his sight? As he wiped his arm across his eyes, he saw it was his own blood that blinded him. He pressed his handkerchief against his forehead.

Terrific, he told himself. If the ocean and the lightning don't get me, I'll bleed to death.

He sat for a moment, confused. Finally, he locked the wheel and made his way below. The pitching boat threw him against the sides of the companion way. With considerable effort, he got below and located the first aid kit. The boat lurched, and he dropped it as he fell across a bunk. As blood ran down his face, he blindly retrieved the contents.

After he applied folded gauze, he tried to secure it with adhesive tape, but the water from his soaked hair ran down his forehead and prevented the tape from sticking. He grabbed a towel and wiped around the wound, wincing from the pain. When he got the skin dry enough, he applied the adhesive again. At least, he hoped, the bandage would stem the flow of blood. He grabbed a hat and jammed it low on his head so that its brim held the bandage in place. Slowly, he struggled back to the wheel. He sat weak, exhausted, and nauseated.

Still dazed, he guided the Columbia through the final agonies of the storm.

Chapter Two

"Cody! Where the hell are you?"

Cody Matheson pressed the mike button. "I'm on the Sebastian River, looking for your marina."

He eased the Columbia's wheel slightly to swing the yacht's bow to starboard as he followed the river's channel. "Continue on up the river, you old fart. I'm just past the boat yard. You can't miss it. I'll be waiting on the dock."

Cody unconsciously touched the soiled bandage on his forehead. The cut was still tender.

After passing through the St. Augustine Bridge of Lions, Cody had motored down the Intracoastal Waterway to the San Sabastian River. Here he turned north, keeping his 32-foot Columbia in the center of the narrow channel.

"And hey," Mitch Larson radioed back, "thanks for coming. Karen with you?"

"I'm alone."

Seated behind the wheel in the Columbia's cockpit, he followed the buoyage system on up the narrow river, red buoys on the right, green on the left, and surveyed the surrounding wetlands. Snowy egrets shuffled their feet to stir up food. Ibis warily watched the boat from where oysters had adapted to the ebb and flow of the tides. Heron fished along the edge of the *Spartina* grass, and an osprey swooped down to catch a fish. A flock of wood storks wheeled against the blue late afternoon sky.

As he approached the Cormorant Marina, he throttled back. Mitch waved him to where he was to moor against the dock; but Cody had trouble concentrating on maneuvering, because of a young woman lying naked on her stomach on the foredeck of the sloop tied up ahead of him.

"Throw me a line," Mitch shouted. The Columbia struck the pier hard. "Jesus Christ," Mitch yelled, "you don't have to wreck the goddamned dock. Where the hell you learn to sail?"

As Cody tossed the line, pain shot through his shoulder. He glanced at the woman, whose expression showed she enjoyed the mess he made of docking.

Mitch secured lines both fore and aft, then straightened and clawing at his crotch, cried, "Welcome to the Cormorant Marina, you old bastard."

He was as usual, unshaven and had on a soiled T-shirt and ragged shorts. Not much different from the way he dressed at graduate school. He was deeply tanned.

Mitch removed his cap, revealing a few wisps of what had been more ample hair, and wiped his forehead with his arm. He had aged.

"You got a broken shroud," Mitch noted. "What happened?"

Cody dropped down onto the dock. "Ran into a storm."

Both embraced and then shook hands.

"Damn, it's good to see you, Cody," Mitch said.

"You too, Mitch."

"And how'd that happen?" Mitch pointed to the bandage.

Cody told him he got hit by the broken shroud.

"You see a doctor about it?"

"They don't make house calls at sea."

"You're a smart ass, Cody." Mitch grabbed both of Cody's arms. "You don't know how glad I am you've come." He hesitated, "I really didn't expect you."

"A lot has happened."

"I want to hear everything. Christ, I'd invite you to stay with me, but all I got is a flat with one bedroom and a kitchen. We got a cafe here; we'll have supper and talk."

Cody gestured at the Columbia. "I got all I need on board."

"We'll get that shroud fixed. And if there's anything else you need, you got it."

"I'll have to scrape and paint the hull."

"I'm at your disposal, Cody. I can haul you tomorrow." Mitch scratched his crotch. "And listen, everything's on me—supplies, docking, everything."

"I appreciate that. The divorce pretty well wiped me out."

"Divorce?"

"I'll tell you about it later."

Cody had noticed the blue and white sheriff's car parked in front of a building he guessed was the marina office. "Trouble?"

Mitch half turned. "That's part of what I wrote you." He paused. "Harassment. I'll fill you in during supper. Look, you old road-kill, I got a boat to haul, but thanks for coming. I mean that."

He left, but not before tossing an appreciating glance at the naked woman, who had gotten up from her sun bathing and dropped down onto the dock.

However, Cody noticed, she wasn't naked. She wore a skin-colored thong.

"Hello." She had her light brown hair cut short and smiled with a mouth Cody thought was too wide, but which he found sensuous. Her deep blue eyes sparkled with excitement. He noted her small waist, well-rounded hips, and good legs.

"I'm Jeanette Severino," she said, then added, "you through taking it all in?"

"That doesn't leave much to the imagination."

"You don't approve?"

"I didn't say that."

"He was right," she remarked.

Cody looked puzzled.

"Your forehead. It's infected, and the bandage is soiled. You better let me tend to it."

"Thanks, but I can take care of it".

"You let it get infected. That's not taking care of it. Come on." She took his arm with a firmness that discouraged objection, and steered him toward her boat.

Below, she brought out a first-aid kit. "Where you headed?"

Cody found the interior of her thirty-foot Down Easter neat and clean. Color photographs, probably of the local area, hung around the cabin. He told her he had reached his destination.

"Why here?"

"I understand Mitch has problems."

She regarded him with mild surprise. "He wrote you about it?"

She laid out the contents of the kit.

"He said something about a developer putting up condos on a wetland and wanting to take over the marina."

"Gear."

"Who?"

"Ed Gear. He's constructing condos on the wetlands across from the marina."

"Why does he want the marina?"

"So the condo owners have a place for their boats. But I suggest you talk to Rocky."

"Rocky?"

"Rocky Brinson. He lives aboard here with his wife. Knows everything that's going on."

She motioned for Cody to sit on a bunk.

He glanced at the pictures stuck up around the cabin. "You're obviously a photographer?"

"I'm heading for the islands, I hope to sell to travel magazines."

She leaned over him as she slowly peeled off a strip of adhesive tape. "Ouch !"

"Hold still. It's stuck."

He enjoyed her close to him.

When she removed the bandage, she applied disinfectant to a cotton swab and dabbed it on the wound. It smarted.

"Hey, he said, where'd you get your training, Auschwitz?"

"Don't get funny." She bent closer to examine the cut.

"Sorry."

"Where'd you know Mitch?" she asked, as she cut gauze into a square.

He gave her a rundown on meeting him at the university. "Mitch didn't finish. He became disillusioned, because professors weren't more involved in the real world."

"What was your major?"

He told her he had a PhD in marine ecology.

Jeanette applied the pad and reached for the adhesive tape. As she did, she glanced down at his ring. "Where's your wife?"

He had forgotten to remove it. "We're divorced."

She gently applied another adhesive strip.

He liked the smell of her. "Where are you from?" he asked.

"I grew up in New Jersey." She put on a final strip of tape. "Heard you tell Mitch you ran into a storm."

"Off the Georgia coast. That's where the shroud snapped." He told her about spraining his shoulder when he went overboard.

Jeanette gave the bandage a final glance. "That cut should have been stitched. I'm afraid you'll have a scar."

"The perils of fighting the sea."

She began putting the first-aid kit away. "You better let me change it in a couple days."

"It feels better already." He gingerly touched the bandage. "Thanks."

"No problem, Professor." She studied him a moment. "That's what I'll call you, Professor."

Cody got up. "Mitch mentioned a cafe."

"It closes at eight. If you like, we'll go up together."

"Better make a lot of noise. I haven't slept much lately," He paused at the companion way. "This developer, Gear, has he threatened Mitch?"

"I think so. In a way, at least. But Mitch isn't about to give up the marina. He and his wife developed it."

"Wife?" Cody questioned. "I didn't know he was married."

"He has a daughter in the deaf and blind school here in St. Augustine."

"Mitch never mentioned a wife and daughter."

"His wife was killed two years after they were married. He never talks about it."

Cody started up the companion way.

"Wait," Jeanette said. She handed him a bottle of liniment for his shoulder.

Back on board the Columbia, Cody applied the liniment and then stretched out on his bunk. Sleep came, but not before, in his hypnogogic state, he reviewed the events that led him to St. Augustine.

CHAPTER THREE

He had often wondered if their marriage was heading for the rocks. Karen was the direct opposite of him. She saw life as a cabaret, and while at first he found this amusing, he began to see nothing amusing when she streaked naked through a home football game, with both teams in hot pursuit. He also wondered if he should go through with the wedding plans when he learned she popped out of a paper mache cake , wearing only a thong, during a faculty stag party.

He reasoned, however, all this would change with marriage, and he was very much in love with her.

While he had hoped there might be a way to salvage their marriage, that notion evaporated the night he and Karen returned from the Friday wine and cheese party at the Virginia Tech University Club.

Cody didn't enjoy these events, but Karen did. Most of the faculty saw them as a chance to complain about students and gossip about who was having an affair with whom. As the wine flowed, tongues loosened and truths became outrageously embellished.

That fateful Friday evening Cody stood on the edge of the crowd, sipping a glass of Chateau Neuf du Pap. The conversation didn't interest him and, ready to leave, he glanced around for Karen. He didn't see her. Probably off talking with a group of faculty wives.

"How goes it, Cody?"

He turned to face Steve Hansbury from the sociology department. Tall and thin almost to the point of emaciation, the man had recently received tenure.

"Anything new on your tenure?" Hansbury asked, once he'd downed a glass of wine.

"Not much," Cody said evasively. He had been thinking of applying at another university. So far he'd said nothing to Karen. Virginia Tech had, in his opinion, become a good ol' boys club, and if you didn't fit in you didn't get tenure.

He saw Karen come into the room. Her face was flushed as she went directly to the wine table.

"Where 've you been?" Cody asked, joining her.

She uttered a quick laugh. "Oh, do I have a surprise for you."

"What is it?" Cody sensed something ominous in her tone.

She threw him her best innocent look. "When we get home, darling."

In their kitchen Cody mixed them each a drink. Then Karen, looking over the rim of her glass pulled a sheet of paper out of her bra. "Look at this."

When he'd read it, he stared at her. "Is this another of your tricks?"

"No."

"It has to be. Raskin would never sign this."

"But, Sweetheart, he did."

"I don't believe it."

He ran his fingers through his hair and finished his bourbon. "You forged his name>"

"If you'd seen him coming all over the place, you wouldn't say that."

He stared at her in disbelief. "Say that again."

"You'll find that's the bastard's signature." She wrapped her arms about his neck. "He never got into me, darling. But we have the memo. You'll get tenure and we stay. And you'll be director of the institute."

Cody was well aware that she wanted to stay in Blacksburg. She had many friends here and was in line to be president of the Faculty Wives Club.

He removed her arms and pushed her back and then slammed his fist on the counter, rattling the salt and pepper shakers.

"What do you mean that he didn't get into you? Holy shit, what in hell happened?"

"I think it's obvious."

"You had sex with the head of the department?"

"Of course not. Not actually."

"What does that mean?" He paced the kitchen floor. "Am I to believe you got the bastard to sign a memo recommending tenure?"

"Yes."

"Oh Christ. Don't you understand. This won't work. Even if it did, what the hell would everyone think? My wife screwed the department head so I could get tenure. Maybe you didn't' actually have sex, but who would believe it?"

He poured another drink.

Karen mixed her own. "Just let me explain what happened."

While he stared at her, she told how she'd suggested she and Raskin go outside to talk. "The little bastard was really horny." She said they went to his car, where she undid her blouse. Raskin wasted no time removing his jacket and pants. Immediately he began to paw her.

"I told him, 'Dr. Raskin, you are very impulsive. Not so fast.' "

"Jesus," Cody groaned.

He was breathing heavily. " 'Golly, you're a desirable woman,' " she mimicked, giggling. "He shifted around in the seat so he could get up against me."

"Damn, have you lost your senses?" Cody shouted.

"I told him this is not something to rush into, no pun intended." She went on to relate that as she slipped off her panties, Raskin was unable to believe his good fortune and groaned, " 'Oh my god. Oh my god.'

"I wanted to make sure he was where I wanted him, so I grabbed his cock. You know what he said? " 'I don't want a hand job.' I told him I wanted him ready."

Cody downed another drink and glared at her.

She explained this his penis was close to entering her as he breathlessly thrust his hips like a rutting goat.

"And that's when I brought out the memo for him to sign recommending you for tenure. He wanted to know what it was and when I told him he said he couldn't sign it. I said I was getting out of the car. But I intimated that if he did sign it, there might be more of what he would be getting."

Cody saw his world collapsing.

"Of course he didn't want to sign it. He wanted to discuss it. 'What is there to discuss?' I asked him. Well, he then said, 'Gee, I don't know. I've never done this before. I mean, under these circumstances.'

"Don't you like the circumstances? I asked."

" 'Of course,' he replied."

Karen finished her drink. "That's when I told him I wouldn't fuck him. The word must have shocked him. Anyway, he signed it. Then the bastard said, 'It better be worth it.' "

She related that Raskin threw himself at her, and as he did, she moved as though to make herself more accessible, but she had maneuvered their bodies so as to put him off balance. When he made his desperate lunge, she thrust her thighs sideways and Raskin's awkward position became his undoing. One knee slipped off the edge of the seat and he fell, semen spurting over his groin and thighs.

" 'Ahhhhh…ohhhhhhhhh…Geeeeez.' " she reiterated.

She explained how she tired to slide back from the source of the sticky ooze ejaculating all over the place.

" 'Damn, damn,' I yelled at him. Give me your handkerchief.'

"As he fumbled for his trousers, he banged his head on the steering wheel. I wiped myself off as best I could."

"Oh my god," Cody groaned as he resumed pacing.

"I told him it was his fault, that he was too damned impetuous."

" 'But I fell off the seat,' he said."

"You had your chance, I replied and got out of the car. Of course I checked to make sure I still had the memo, and headed for the ladies room."

"Goddamn!" Cody yelled.

"But nothing happened, darling. What's the harm? My plan worked perfectly."

"Thank god for little favors," Cody shouted. "Who would ever believe it?"

"My sweetheart, who cares?"

"I do. Now tell me, what in god's name am I supposed to do with this?" He picked up the memo and waved it at her. "Hand carry it to the tenure committee and tell them that Raskin signed it so he could make love to my wife?"

"Send it through the campus mail."

"No."

"Why not?"

"I wouldn't accept tenure after this. Besides, I've made up my mind."

She regarded him questioningly. "Oh?"

"I'm leaving Tech."

Her mouth dropped open. "You're what?"

Cody didn't object to the divorce settlement in which she got the banking account, the house, and the car. He simply wanted out. She didn't, however, get the Columbia, because she hated sailing.

The fact is, Cody reflected just before sleep overtook him, he wasn't sure just what she liked, except to engage in those crazy antics. At first, they rather intrigued him, when they, met at the University of Maine, where he was finishing his doctorate and where he also met Mitch Larson. Karen was working on a master's in art. Cody was drawn to her vivacious dark beauty and her flamboyancy. He never thought during their whirlwind courtship that they weren't suited for one another.

During the divorce proceedings, Cody moved onto the Columbia in a Norfolk marina. The letter from Mitch drew immediate anger. A developer was planning to built condominiums on a pristine wetland across the road from Mitch's marina. This was bad enough, Mitch wrote, but the developer wanted Mitch's marina and threatened to get it one way or the other. Mitch concluded by writing that the area was beautiful and unspoiled. Mitch was as avid an ecologist as Cody.

Mitch apologized for dumping his troubles on Cody. "I know you have your job and marriage to think about, but I have to unload on someone." He didn't know, of course, about Cody's situation.

As he finished the letter, Cody reflected on their graduate school days. Mitch had worked toward a doctorate in marine ecology. He had been an average-looking guy with intense blue eyes and thinning black hair. Unless he had to dress up, he wore tank tops and ragged shorts. Mitch habitually called people he liked by uncomplimentary names and occasionally referred to Cody as a fructifying road kill. Cody's Nordic features and blond hair were in stark contrast to Mitch's swarthy appearance. At six one, Cody was several inches taller than Mitch.

Mitch didn't finish his doctoral program. A third of the way through he declared he was fed up with academia, the hypocrisy of the professors' preaching about saving the environment, but doing nothing about it, exasperated him.

"What will you accomplish by leaving?" Cody asked.

"Peace of mind."

Here days later Mitch put his what money he had into a 30-ft Morgan sloop and sailed south. He sailed for a year, sending postcards to Cody from Martinique, Jamaica, St. Thomas, San Juan, and Grenada. Finally, he wrote to tell Cody he sold the boat in St. Augustine and bought a marina, which he named the Cormorant Marina. He added that Cody would always be welcome.

Cody and Karen were married when Cody finished his course work and accepted a research position at Woods Hole on Cape Cod. A few years later he took the teaching and research job at Virginia Tech.

CHAPTER FOUR

Shortly before Cody arrived in St. Augustine, a red Thunderbird turned down Riberia Street. The driver looked in the rear-view mirror. "What's the status of the Cormorant Marina, Sid?"

From the cramped rear seat, the vice-president for real estate stared at the back of the driver's head, not wanting to make eye contact in the mirror. "Doesn't look like Larson'll sell."

The president of Condo Developer, Inc. glanced at a group of blacks sitting in the shade of a building. The area was Lincolnville, south of downtown St. Augustine that had been established after the Civil War for freed slaves. Now, mostly blacks, but a few whites, lived here in old Victorian homes, shotgun houses, and cinderblock structures.

"Goddamnit, Sid we got to have that marina," Ed Gear snapped. "Get off you damned ass and do what you have to in order to make him sell."

Gear slowed as he approached an unpaved road that led into a large wetland with areas of pines, cypress, oaks, and palmetto. As he turned sharply into the area, he glanced to his right at the entrance to the Cormorant Marina across the street. One way or the other that marina would be his.

He switched off the engine and unfolded from behind the wheel. Dark glasses hid his blue eyes, but his thin mouth and dour demeanor suggested a man who was used to getting his own way. He wore light blue slacks and a short-sleeved sport shirt. The others were dressed similarly.

Taking the lead, Gear strode to the edge of the wetlands and looked out across it. The others followed. Sidney Phillips wiped his cherubic face and the back of his neck. He wheezed from the exertion of walking. Juan Cordover, vice president for development, stood next to Gear. He was shorter, with swarthy features, and more solidly built.

Gear hated the smell of salt water, decaying vegetable and marine life. He shook his head with disgust. "Got to do something about the god-damned smell." His comment was directed to no one in particular.

He turned to Phillips. "What about the property along Riberia Street?"

"It's mostly owned by niggers and a few whites. I have options on almost all the tracts. I should have them all wrapped up within a couple months."

Gear glared at him. "A deal isn't a deal until its signed, damnit. We got to have the land so we can widen the street leading into the condos."

"Right," Phillips replied. He knew he'd given the wrong answer.

Gear took hold of each man's elbow. "What do you guys see?"

"A swamp," Phillips replied. Cordover was smart enough to wait to see what his boss was getting at.

"No, Sid, that's not what you see." He released Sid's elbow and made a wide sweep with his arm. "You see condos, tennis courts, landscaped lawns, putting greens, a snack bar, a shopping center, and parking lots."

He moved away and stood with his hands on his hips, as he saw the completed project.

"The entrance from the street will run along there," He pointed to where they had entered the area. He looked at Cordova. "Start clearing the trees and underbrush. We don't have the permits, but fuck that. We can't wait for the agencies to get off their fat asses.'

Before returning to the car, Gear glanced out across the wetlands. He felt good about the purchase of this 200 acres. He'd bought it from the city of St. Augustine at a fraction of its market value. The city didn't want it and didn't know what to do with it. In buying it, Gear had mort-gaged himself and the company heavily. But he was pleased that his wife, Erica, liked the deal. She had a good head on her shoulders. And both saw it as one of their most profitable investments.

Gear smiled, recalling that the city commissioners hadn't realized the potential value of the purchase. To them it was just a stinking swamp and were delighted to find someone crazy enough to buy it. It had been given to the city by a widow with the stipulation that its natural ecology would not be destroyed. Gear knew this. The mayor and the commissioners knew this. No one saw any reason to bring it up during the negotiations.

To make the development profitable, Gear would have to acquire the Cormorant Marina. The marina would give the condo owners access to the intracoastal waterway and a mooring area for their boats. Once he got the marina, he planned to expand into the adjacent wetlands. Such an expansion would require the approval of the State Department of Environmental Regulation, but he was confident he could get around the regulations. Also, because the area where he planned construction received tidal flow from the Matanzas River, he'd need the Corps of Engineers approval for dredging. Gear had a plan for getting that, too.

"We've got too many goddamn rules and regulations," he muttered as they returned to the Thunderbird.

Nevertheless, there were ways to circumvent those, and as he turned on the ignition, he congratulated himself for having the foresight to enlist the support of State Senator Victor Blanchard, a powerful political figure on the state appropriations committee. Blanchard promised to negotiate with the Department of environmental Regulations to facilitate getting the construction permits. He also put a considerable amount of money into the project, for which Gear offered him a couple condos. These would bring in a tidy sum when the project was completed. Blanchard had fairly salivated over the prospect of reaping a handsome return for his investment.

As they drove out of the area, Gear reflected on the initial $2 million line of credit he'd obtained for the architectural plans, legal fee, buy-outs, and, of course, payoffs to local officials. But this amount was just the beginning. He'd need much more before the project was completed.

It didn't escape him that there might be local opposition to the condos, especially from environmentalists. Fortunately, however, these groups were not well organized, and could not muster a strong front against the project.

Gear had inflated the amount of money the project would bring into the city as well as the number of jobs it would provide. That was the way of doing business. A few hints that some of the condos might be available to the mayor, influential business men, and the city commissioners, ensured there would be no objections from these people regarding harm that might come to the environment from the project.

As the Thunderbird swung out of the site, Gear glanced again at the Cormorant Marina. A weathered sign erected over a small building indicated a cafe. Well, he'd get rid of that, and imagined the changes he'd make once the marina was his.

Gear was eager to see the entire condo project completed, not only for the money, but the prestige he would derive from putting together such a magnificent project. It would be a complex like nothing else in Northern Florida.

Ten minutes later Gear turned off San Marco Street into a parking lot behind what had once been an elegant three-story home and was now the offices of Condo Developer, Inc.

Inside, Gear looked in on his wife, as Cordova and Phillips headed for their offices. She sat behind a large mahogany desk. Her blonde hair swept back in natural waves reflected golden highlights from the late afternoon sun coming through the window behind her. Her deep blue discerning eyes greeting her husband with an almost expressionless stare.

"I'm back," Gear declared.

"So I see."

She got up and followed him into his office.

"I've been thinking," she said, sitting down.

Behind his desk, Gear leaned back with his hands clasped behind his head. "Oh? About what?"

She crossed her legs, revealing an expanse of thigh. He was immediately aroused. The first time he saw her at the Boston Chamber of Commerce party over six years ago, he couldn't keep his eyes off her. Tonight he'd take her to supper and after a few drinks at home they'd make love.

"I want to be sure we've covered all angles," she said, giving him a look that all but dismissed what he had in mind. He hated it when she did this.

His thoughts went back to the Chamber of Commerce party. He recalled she was there to do a short TV piece for the station she worked for. After introducing himself, he hinted that perhaps there could be a story about his company. A construction firm that put up condominiums throughout New England. She seemed mildly interested and told him she'd speak to her boss. Whether she did or not, Gear was never contacted. A week later he phoned the station and asked her out for dinner. At first she seemed reluctant, but he persevered.

Although he took her to a fashionable restaurant, he felt their first date a disaster, mainly because he spent most of it talking about himself. How he'd worked his way up in the construction business, which wasn't bad for a kid raised in the Boston slums. However, she went out with him again, and this time he let her do the talking.

He immediately saw she would be an asset to his firm and offered her a job. She wasn't, she explained, ready to give up her present position. She had her eye on being anchorwoman. Gear let it go at that, but brought the subject up on a later date. This time he suggested she could be a senior executive. The pay would be very good. Not until he suggested she would eventually become a partner that she relented and a month later joined the firm.

"There's nothing to worry about, sweetheart," he replied to her concern. "I've been making the right contacts to grease things along, as I'm sure you're aware.

"And the environmentalists?"

He dismissed them with a wave of his hand, thinking how beautiful she was, and as his mind blazed with thoughts of sex with her.

While his first reaction was to say the hell with the environmentalist, he didn't want to put her on the defensive. "Look," he leaned forward on his elbows. "I've talked with Ken Johnson over at the Sierra Club and Burt Higgins with the Audubon people. Both like our plan. They see it as a boost to tourism and greater prosperity for St. Augustine." He smiled. "Of course they jumped at the idea I might donate some land for a wildlife preserve. And I've spoken with the head of the Wilderness Society, Ray Coontz. He had no objections to our condos, especially after I suggested he might be in line to get one.

"Those guys spend more time worrying about budgets, appropriations, and looking after their fat Asses. And as for the environmentalist, they aren't organized here in St. Augustine."

Despite his lack of prowess in bed, Erica had to admit her husband had a way of handling things,

"As you know," he went on, referring to the sheriff and chief of police "I've talked with Banks and Riker. They assured me they wouldn't put up with any nonsense from those kooks who go around worrying about the ecology. I tactfully reminded them we contribute to their organizations."

"Very good, but we can't afford any loose ends. What about the lab?"

The lab was Erica's idea. She had suggested that it be located at Flagler College, a small undistinguished liberal arts institution.

She had explained, "If we could get it set up as both a teaching lab and a lab to conduct environmental assessments on properties that we plan to develop, it's just possible that would satisfy the state and federal agencies making the assessments, so they wouldn't have to do them. I'm sure you see my point, Ed," she added.

He liked the idea, very much, in fact. The lab would keep the permit granting agencies out of their hair. "We have to convince Flagler," he told her.

"Will that be so difficult? We offer to pay for part of the salary of the professor who's assigned to run the lab—one who'll see things our way—and make them see the lab as providing training for students majoring in biology and ecology. The lab could be a prestigious addition to the college."

Gear recalled with amusement the eagerness with which the administrators grabbed at the idea. At first they did not appear too eager, a ploy to put them in a better bargaining position, perhaps holding out for the Gears to pay the full salary of the professor. However, within a week the administrators agreed to all terms, and suggested a Dr. Rod Chapman as the lab director.

Ed and Erica interviewed Dr. Chapman and agreed he was a wise choice. A small, man with am ingratiating demeanor, he appeared one whom the Gear's could easily manipulate. He was approaching retirement, which he didn't want to jeopardize.

As Erica left his office, Ed Gear leaned back. He had to make two phone calls. One was to Senator Blanchard and the other to Bill Gooney, head of the Corps of Engineers. Gear had something on Gooney. Now he would use it.

CHAPTER FIVE

He dreamed someone called him.

Cody opened his eyes and looked around the Columbia's cabin.

"Professor."

He got up groggily and looked out the companionway. Jeanette waited on the dock.

"Are you going to eat, Professor?"

He would rather sleep, but he was hungry. "Give me a minute," he muttered.

After splashing water on his face from the galley sink, he put on clean slacks and shirt.

As he dropped onto the dock, Jeanette glanced at his bandaged forehead. "How's that cut on your forehead?"

"It's fine, thanks to you."

On the walk through the boat yard, Cody got a better picture of the layout of the marina. A pier extended straight out from the shore. Two docks extended at right angles from it. Cody's Columbia was moored on the outer one.

He glanced at a grassy area to the his left where there were several picnic tables and a couple fifty-gallon drums cut in half to make a barbecue grill.

To his right stood a wooden shack with rails extending down into the water. A thick cable, used to haul larger boats out of the water that could not be lifted by the mobile crane, ran between them.

He watched a flock of wood storks wheel against the early evening sky and a kingfisher dart across the wetland. A blue heron stalked its meal in the shallows. The pungent odor of decaying vegetation and sea life rose in the wake of the ebbing tide

"Interesting." he remarked as he nodded with amusement at the names painted on the sterns of the dry-docked sloops, ketches, and yawls, and an occasional power boat—Unnatural Love, Escapade, Ophelia Fanny.

"They look out of place," he remarked, "like sea creatures that have crawled out of the ocean on iron legs."

"Very descriptive, Professor."

"I guess you get people from all walks of life here."

"Pretty much," Jeanette replied. "Some have a lot of money. Some don't. With boaters it doesn't matter how wealthy you are."

"That's good, because my finances aren't that great."

"The divorce?"

"She got everything but the boat."

"I'm sorry."

He dismissed her concern with a wave of his hand. "I didn't want any hassle."

"She must not have been a nice person,"

Before he could answer a man emerged from behind a dry-docked yawl.

"Hello Ben," Jeanette greeted him. "How's the shrimping?"

He stared at Cody. "It's okay. I got in this morning." Still looking at Cody, he asked, "Who's this guy?" Cody stared at his wide, dark face and brooding eyes.

"This is Dr. Matheson."

"Just got here, huh?" Ben remarked.

"That's right."

"Plan on staying long?"

Cody did not like his tone. "It depends."

"On what?"

"Like I said: it depends."

"Ben," Jeanette cut in, "he's a friend."

"Yeah?" Ben continued to stare at Cody. He turned to Jeanette. "I thought we might get together, now I'm back."

"I'm sorry."

Disappointment camped on his face. "You like this guy?"

Cody lost his patience. "That's none of your business."

Ben took Jeanette's arm and began to lead her away. "We have to talk."

"Stop it, Ben," she demanded. "Let's not have any trouble."

"We got things to settle," Ben said holding her arm.

Cody looked Ben directly in the eye. "There's no need of this."

Ben let go of Jeanette. As he did, he swung around and threw a punch, which caught Cody on the side of his jaw. The weight of his muscular body behind it, the blow knocked Cody backwards.

A stab of pain shot through Cody's shoulder as he hit the ground.

"Stop it. That's enough, Ben. Stop it!" Jeanette stepped between them. "If I report this to Mitch, he'll throw you out of the marina."

As Cody got up, Ben said, "We got to talk."

"Ben," Jeanette replied, trying to reason with him, "I think we just better leave things as they are."

Ben studied her face. "We'll talk later."

Jeanette put her hands up to signify she didn't want to pursue this further. "No, Ben. We have nothing to talk about."

She took Cody's arm.

"Why won't you talk with me?" Ben pleaded.

"Ben, it just wasn't for us to get together. Please understand that."

"What have I done?'

"You've done nothing, Ben. Now, we're going to have supper."

As they left Ben, Cody noted Ben hadn't moved, but stood watching them. He felt of his jaw. It was sore but not broken.

"You all right?" Jeanette asked.

"He throws a mean punch," Cody said. "What's with him?"

"He's Ben Hanks. Works on a shrimp boat. He's got the hots for me and thinks if he keeps pestering me I'll go out with him. I have no intention of doing so. But I do feel sorry for him."

"How so?"

"Well, he hasn't many friends."

"I can understand why."

Cody noticed the cafe's sign had seen better days. Inside, he smelled the pleasant aroma of food, which reminded him he was hungry.

His mind traced the delectable smells back through the swinging lattice doors that led to imagined simmering pots, sizzling fry pans, and aromatic grills. He wondered, looking around the restaurant, how sterile the kitchen was.

He glanced at the metal tables, distributed haphazardly about the small room, and the few customers. Cody noticed the floor needed sweeping and the walls needed painting.

Jeanette saw Cody's expression." The food is very good, Professor."

"I'll take your word for it."

Mitch sat alone at a table against a wall.

Mitch noticed the bruise on Cody's jaw. "What happened?"

"We had trouble with Ben," Jeanette said.

"What kind of trouble?"

"He punched the Professor."

"The sonofabitch. I'll throw him out. We don't want that kind of behavior around here,"

Cody held up a hand. "No, let's forget it. Nothing serious."

The lattice doors to the kitchen flew open and an apron-clad man came toward them,

Jeanette explained, "He owns the place. We call him The Chef."

Cody figured the man was in his fifties; tall, with a pot belly. He scratched the stubble on his face. A few thin strands of dark hair lay across the top of his head. He looked like something discarded, a man washed up on the barren beaches of life. He waited, mouthing a toothpick.

Jeanette ordered fried scallops, baked potato, and iced tea. As the chef wrote, spittle ran from the corner of his mouth, which didn't enhance Cody's estimation about the place. "I'll have the fried oysters." Cody said, dubiously.

"What you want with them?"

"How about mashed potatoes and coffee?"

The chef looked down at Cody. "Don't ask me. It's your order. You want 'em you got'em."

"Sounds okay."

Mitch had already ordered.

The chef went back to his kitchen.

Jeanette remarked, "No one knows where he came from."

"He showed up about a year ago," Mitch added, "and offered to set up this cafe. Why not, I thought, as long as I get my cut."

Cody looked at Mitch. "So you got married and had a child, I hear. You never mentioned that in your letters."

Mitch shrugged, suggesting various possible replies. Then he said, "She's doing okay over at the deaf and blind school."

Cody noticed a man who looked in his seventies sitting with an attractive blond, obviously much younger.

"She used to be a stripper," Jeanette explained, "Her name's Lori. The guy's Larry Mackintosh."

"Avoid getting in a conversation with him," Mitch advised, grinning. He didn't elaborate. Instead, he asked, "What made you leave the university, Cody."

"A couple things. One was the divorce. I won't go into that. The other was I knew I wouldn't get tenure for another year, if at all. And I wasn't going to be director of an environmental research lab, as promised, at least not until funding became available."

The chef brought the iced tea and coffee.

"My last interview on tenure with Raskin, the department head, convinced me I had no future at the university. Raskin claimed the university was concerned because I had published in several radical environmental magazines. Raskin wasn't happy about it, either.

"Anyway, he said until I published in respectable journals, the university couldn't consider me for tenure or set me up as director of a research lab. Translated, that means I was rocking the boat insisting we do everything we could to save our wetlands. It came down to politics and money."

Jeanette listened intently.

"The land developers are a powerful group and they exert their influence in congress to water down attempts to prevent construction on wetlands. Land is getting scarce, and so the wetlands are a prime source for putting up housing developments, shopping centers, and whatever, especially here in Florida.

"Because of lobbying and the land developers' pull with congress, research grants went to colleges and universities where the faculty didn't oppose wetland development. The editors of research journals, of course, saw what was happening and favored articles from authors who mouthed the conventional wisdom of the power blocs—in other words, don't push for "no net loss of wetlands."

"The damn bastards," Mitch muttered.

"So I turned to the radical journals to get my message out. The university didn't like it."

"Of course not," Mitch put in.

"When I asked why in hell the university wasn't concerned about the environment, Raskin waved my question away. He claimed we shouldn't get involved. I told him we should get involved instead of sitting around on our asses while our valuable wetlands disappeared. He got pissed off and said he didn't like the way I talked. I said he wouldn't have to put up with it. I quit."

Cody finished as the chef brought their supper. "Exactly what's the situation here, Mitch?"

"Ed Gear is the developer whose putting up the condos." He nodded to indicate the area across from the marina. "He wants the marina so those who buy condos will have a place for their boats."

He paused to sip his coffee.

"Gear has the power structure behind him. That includes the city attorney, the city commissioners, the city manager, and wealthy businessman, they all expect to profit from the condominiums in one way or another. A guy named Phillips, vice-president for real estate, has been out here trying to get me to sell. Same old thing: now's the time to get a good price for the marina. Occasionally his offer is couched in veiled threats, which I laugh at."

"Like what?" Cody asked.

Mitch leaned back. "Nothing concrete. He intimates things might not go to well for me. Gear's done the same thing. Calls and makes an offer, then follows up with the threat he's going to get it one way or the other."

The chef brought their meals. As Cody took his first bite, he glanced at Jeanette, but before he could comment that the food was excellent, a large man, followed by a tall, thin woman. burst into the cafe.

"Hello folks," the man bellowed.

Jeanette leaned toward Cody. "That's Rocky and Millie Brinson. He's the one I mentioned you should talk to."

Brinson had only a few strands of white hair on the top of his head. Blotches of paint stained his T-shirt and pants. The tall woman with him had light, sun-bleached hair and sharp, but pleasant features. Paint spattered her clothes also. They seemed unsure where to sit.

"Why don't we invite them over?" Jeanette suggested.

Mitch beckoned and, when they sat down, Jeanette made the introductions.

"My pleasure, Doctor," Millie Brinson said. Cody liked her warm smile. "What kind of practice do you have?"

Cody told them he had a PhD in marine ecology. "And please," he added, "I prefer Cody."

Millie smiled. "I like that name."

"Millie, here, graduated from Purdue," Rocky injected. "I graduated from the University of Illinois. Majored in engineering." He paused. "We both got fed up with the rat race and took early retirement so we could enjoy the life of wanderers."

He turned to Mitch. "How goes it, my friend?"

Mitch gestured with his fork, swallowed, and said nothing had changed.

"You gonna order?" the chef grumbled, looking at the Brinsons. He had ambled over unnoticed and waited with his pad in hand. He stared at them with vacant, rheumy eyes.

Rocky selected the fried oysters and ice tea and Millie the seafood salad and coffee.

Glancing at the bandage, Millie asked, "Whatever happened to you, Cody?"

"I had a confrontation with the end of a shroud."

"You're damn lucky you didn't get hit in the eye," Rocky remarked.

"I have Jeanette to thank for this." He pointed at the bandage.

"And that bruise on your face?" Millie asked.

"Ran into Ben," Jeanette explained.

Rocky looked at Mitch. "Trouble?"

Jeanette explained what had happened.

Millie was looking at Cody. "Mind if I ask you where you're from?"

Cody told her he'd left Virginia Tech.

"Is there a Mrs. Matheson?" Millie asked. "Or am I being too nosey?"

"Not at all. We're divorced."

He turned to Mitch. "I saw a police car parked by the office when I docked, You said something about harassment."

"As Mitch, here, will tell you," Rocky cut in, "the local cops and sheriff come around now and then. Both are in Gear's pocket. They claim people at the marina pull up signs and survey stakes over at the condo site. None have, least not that I know of, but it's an excuse to keep pressure on Mitch. Gear wants this place in the worse way." He leaned back and crossed his gnarled hands over his large belly covered by the soiled and paint-stained T-shirt."

"Gear," Mitch added, "has built developments in New England, the West Coast, Cabo San Lucus. The sonofabitch bought the land opposite the marina. Got it for a good price, because the town didn't know what to do with it. Anyway, the wetland is partially wooded and extends across the peninsula to the Matanzas River, which forms part of the Intracoastal."

"The area's a natural habitat," Millie cut in.

Cody's mind flashed back to the marsh of his youth and the old man who had lived there.

The chef interrupted with the Brinson's order.

Rocky looked down at his oysters with obvious relish, then glanced appreciatively at his wife's seafood salad. "Can you imagine," he said in a good-natured way, "that she's watching her weight."

If anything, Cody thought, she could use a few more pounds, but he was more interested in what Millie had said about a natural habitat.

Rocky forked an oyster. "Damn, these look good."

"How's your salad?" Jeanette asked Millie.

"Couldn't be better," Millie talked and chewed at the same time.

Cody pressed for more information. "If Gear is constructing condominiums over a natural habitat, is anything being done to stop him?"

Rocky took a drink of his iced tea. "You'd think something could be done, but you got to know about politics around here. It comes down to a matter of clout. Gear's got a state senator on his side. The condos mean jobs and money. Rumor also has it Gear isn't the only one who's invested in the project. If that's true, many people have a strong interest in seeing it succeed."

As Rocky opened his mouth for another oyster, Mitch took over, "He's got the town sold on the condos, he and his wife."

"What about his wife?" Cody asked.

Rocky swallowed an oyster. "Word is he met her when she was working for a TV station in New England. She's a beautiful woman. And smart. She probably has as much to say about the running of the company as Ed. Maybe more."

Rocky paused to enjoy another oyster.

"Getting back to your question, Cody. We both have written letters to the editor of the local paper and to our representatives in Tallahassee. The result? Nothing."

"And who else?" Rocky asked, looking at Mitch.

"Well, I wrote to The Department of Environmental Regulation, the Corps of Engineers, the Chamber of Commerce."

Rocky injected, "Soon after we stared writing, the incidents began."

"Incidents?" Cody asked.

"I've had some equipment stolen. Found the leg of a boat stand cut nearly in half. But the worse thing was someone cut partway through a link in a chain used for hauling out the larger boats. Fortunately no one was hurt seriously, when it snapped."

"And you figure Gear was behind it?" Cody asked.

Mitch looked directly at Cody. "I'd bet on it. However, what we think and what we can prove are two different things."

Rocky said, "We don't like it, although we don't live here, except for short periods of time." He glanced at Mitch. "Mitch is an honest guy. His prices are fair. That's why we come back."

"How about getting an injunction against the project?" Cody asked.

"I can't answer that," Rocky replied.

"What about environmentalists?"

Rocky shrugged. "There's no concerted environmental movement around here."

"If he doesn't get this marina," Millie injected, "he won't have access to the water way. The Corps of Engineer won't let him dredge a channel into the wetland area to the intracoastal. Something to do with drainage and flooding, but primarily because the intracoastal is too shallow near his building site and he'd have to dredge too far out."

Cody thought a moment. "If he couldn't build the condos, then he wouldn't want the marina."

Rocky eyed another succulent oyster. "That might be true, but how's he going to be stopped building the condos?"

Cody resumed eating, as he added, "It infuriates me that our wetlands are being turned into roads, shopping centers, homes, football stadiums, and trash dumps."

"Amen," Mitch grunted.

Jeanette turned to Cody. "I don't want to sound dumb, but why all the concern over what looks like just smelly old marshes?".

Cody finished what was in his mouth. "Fish caught by commercial fishermen need wetlands to survive. Also a great many birds need wetlands for sanctuary as well as feeding grounds. That's just for openers."

He was interrupted as a well-dressed couple entered the cafe.

"That's Tony Pazzio and his wife." Mitch explained. "No one knows much about them. Rumor has it that he was with the Mafia. Maybe he was; maybe he wasn't, but he's got a nice collection of pistols."

Cody noted that Pazzio looked his way and stared at the table a moment. He wore his black hair combed straight back. He had on dark slacks and a light short-sleeved shirt with a cravat. His wife in gray slacks and a pink blouse had a smooth look about her with her pageboy hair. Both looked like they'd just posed for an old movie poster.

Millie turned back to eating. "You may wonder why people here are concerned about what Gear is doing, since we're all just passing through. But we respect the environment. We live in it, more than most people. And we would like to come back here to work on our boats and renew friendships. We don't want to see anything happen to change this place."

As she finished, Rocky leaned forward. "That's Marion Talbert." He nodded toward an attractive, plain-looking woman who had just sat down. "A lesbian." He made air quotes around the word. "At least that's what people say." She's a damned good sailor. Been all over the world."

"We get people from all walks of life," Millie continued. "Bankers, retired businessmen, people searching for their lost lives, all sharing a mutual interest in the sea—along with their idiosyncrasies."

Jeanette smiled. "A regular personality zoo."

"It seems that way," Cody remarked, but he hadn't sailed long enough to form opinions about people who spent their lives aboard boats. Maybe they were a little different. However, he figured it took a special kind of person to sail the oceans of the world.

He looked up to see Larry Mackintosh and Lori approaching. She did have a very good figure. The man walked with an osteopathic stoop. His dark eyes glowed with a wild look. When he stopped at the table, he introduced himself and Lori.

"Haven't seen you before," he said to Cody, as he thrust out his hand. "Must have just got in. I mean, we believe you'll like it here in this corner of life. If that leads to eternal wisdom, so much the better. Of course we need some degree of consensus, because the road into the sky is wide and winding, like the winds across the prairies where the wild wolf calls." He took a breath, girding himself to continue.

"Come on, Larry." Lori took his arm. "Let these people eat. I'm glad to meet you," she said, smiling at Cody. She led Mackintosh away.

"Like we said, we get all kinds," Rocky remarked with a grin,

The group talked for several minutes, and finally Rocky said he and Millie had to get back to their boat.

"We'll talk some more, Cody," Mitch said as Cody and Jeanette got up.

"I look forward to it."

Mitch said he was going to have another beer and go home.

At the door, the four met a short man with a dark beard, long black unkempt hair. He held a can of beer. Rocky introduced Water Rat. The man looked up at Cody with bloodshot eyes and shook hands.

Outside Rocky explained that Water Rat lived on a small houseboat and earned his beer money working on shrimpers. "His real name is Fabius Thornton. I think I like 'Water Rat' better," he added, grinning. "Apparently he's got a mother in Ohio, but no other family we know of. He's okay, except when he drinks." He chuckled. "Hell, I guess all sailors drink too much."

As they walked through the darkness, Cody remarked that he didn't like what he had heard back there. "I knew from Mitch's letter there were problems, but he didn't go into details."

"Mitch isn't a complainer," Rocky replied, "but I'm sure you know that. Anyway, I don't envy him, up against Gear and his crowd."

They paused by the Brinson's 30-foot sloop, with "Flight of Fancy," painted on the stern. Rocky laid his hand on Cody's shoulder. "Nice meeting you, Cody. See you tomorrow." Millie smiled, patted his arm, and said she was delighted to have met him.

Cody and Jeanette returned to their boats.

"It's beautiful," she remarked looking out across the dark wetlands.

The night hung still and the stars hung close. The lights of St. Augustine reflected on the water.

"But threatened," Cody said bitterly.

He was silent a moment. "On my trip down I saw a lot of wetlands being taken over. People don't realize how valuable they are to our ecosystem."

"What are a few acres of wetland matter, Professor?"

"Wetlands help control pollution as they absorb natural chemicals and recycle them to support wildlife. Also they're a primary breeding grounds for many crucial species along the food chain."

Jeanette listened intently.

"The sad thing is they're not seen as an integral part of the entire eco-logical system. We think they're stinking places not good for anything except to build on."

"Maybe you should write a book, Professor."

Well, he thought, that might be something to consider.

For a moment longer they stood looking out across the dark wet-lands. Cody turned to her. "How about joining me for some wine?" She smiled. "Okay."

Aboard the Columbia, Cody went below and returned with two glasses and a bottle of wine.

"But you have to admit, Professor, the wetlands do smell."

"Mostly when the tide is out, like now."

"I could do without it."

"Despite the odor, wetlands are important. If you destroy a wetland you affect a lot more than just that one area. Like the ripples moving across a pond when you toss a stone into it."

Cody poured the wine. "You haven't told me much about yourself except that you're from New Jersey," he said.

She looked into the darkness toward St. Augustine. "Have you heard of Vito Severino?"

He hadn't.

"Perhaps you wouldn't, unless you're up on the Mafia. Vito is my father. My mother doesn't approve of what Dad's involved with, but she can't do anything about it." She paused. "I decided I could. I pulled up stakes. Mom and Dad gave me the money to buy my boat. And so here I am."

"You still in contact with your parents?"

"Off and on." Then she asked, "Do you like to dance?"

"Yes, but I may be out of practice."

"There's a combo every Saturday night at the Palms. It's a place north of town. Would you go with me?"

"Why not?"

She smiled at him. "It's a date. By the way, if you need to get into town, I have a car. It's an old Chevy. A guy left it to me before he sailed."

"I may take you up on it."

They talked until the wine was gone.

Finally Jeanette said, "It's been very nice, Professor. I better get back."

After going below, he lay naked on his bunk and reviewed the events of the day. Jeanette was certainly an attractive woman. She was considerably different from Karen. He didn't think Jeanette would pop naked out of a box at a faculty stag party or run naked down a football field.

CHAPTER SIX

After Mitch hauled the Columbia, Cody spent two days sanding the hull. A fine blue dust covered him. Rivulets of blue sweat streaked his face and naked torso. His hair resembled a botched dye job. And then he began painting. At the end of each day he looked like a circus clown someone had rakishly splattered with blue paint

When finished, he stood back and surveyed his work.

"Looks good." Gus Johnson had approached unnoticed.

"Thanks."

Cody put the roller in its pan and wiped his hands on his paint-stained shorts. He wondered, glancing at Gus's sloping shoulders and big arms, if the guy had been a boxer during his military career. A big man, Gus still hadn't gone soft in the middle. However, tough jarhead or not, word around the marina was that his wife dominated him, but no one mentioned that to Gus.

"Looks like you got more on you than on the boat." Gus chuckled, as he rubbed the stubble on his lantern jaw.

Gus wanted to chat, but his wife's strident voice brought a grimace to his swarthy features. "You let me know if I can help."

But before he took off, he told Cody he'd heard about the trouble with Ben. "Listen, I'd like nothing better than to cold-cock that bastard. You say the word and I'll see he doesn't bother you again."

"I appreciate that, Gus, but I don't want to stir up any more trouble."

Gus took off at double time,

During the week the sound of bulldozers and chain saws grated on Cody's nerves. The Gear's had wasting no time clearing the land in preparation for the condos. Cody tried to ignore the sounds, but the deep-throated growl of laboring engines and the strident whine of chain saws were not easily put out of mind.

A large shadow fell over him. He turned to face Jugs who stood watching him, as she juggled six tennis-sized balls.

"You doing okay, Cody?" She seemed oblivious to her juggling act.

He knew she had been with a circus and that her real name was Clara. Whether the sobriquet 'Jugs' was a shortened version of 'juggler,' or a description of her immense bosom, Cody was not sure. He liked her husband, Jake, a stocky, muscular man, who loved to talk about his experiences as a circus roustabout.

"I'm finished," he said.

She opened a bag tied at her waist and let the balls fall in one by one.

"I'm impressed," he told her.

She laughed heartily; her massive flesh quivered. "You ain't seen anything once I could juggle ten." Her dark eyes sparkled. "There's a lot you don't know about me, Professor." Jeanette's nickname had caught on.

He should have been prepared for her powerful jab, which was her way of greeting. He'd experienced it before. But this time he was caught off guard, and reeled backwards, stepped into the pan of paint, slipped, and fell against the wet hull. He came away with his hands, chest, and face deep blue.

"Geez," she exclaimed, "you're a mess. We ought to get up an act."

He wouldn't encourage that idea, he thought, as he wiped his face with the paint-stained rag and smiled ruefully. Jugs broke into laughter.

"See you later," she said, and walked away, chuckling, leaving him looking like a painted harlequin. He watched the huge cheeks of her posterior performing their own juggling act.

Cody cleaned his equipment and put it away, then climbed aboard and put on a bathrobe. With a towel, a bar of soap, and a can of turpentine, he headed for the men's shower.

Outside the bathrooms he found Jeanette in a bathrobe, waiting for the women's room to be available.

"Hi, Professor."

He glanced at the western sky. The sun had set and a brilliant pink colored the last of the building thunderheads. A stillness had settled, as the sea breeze slackened, waiting to reverse course and become a land breeze.

The door of the men's room swung open and Water Rat stumbled out and staggered down the steps with a can of beer in one hand. He glanced at Cody, waved, and continued, leaving the shower door open..

Cody knew that people at the marina shared both facilities. "After you," he told her, nodding toward the men's bathroom.

Jeanette smiled. "That's nice of you, professor. But why don't we both use it? You're a mess with that paint all over you. You'll need help getting it out of your hair."

"I don't mind waiting."

"Don't be a prude." Everyone, married couples, those shacking up, they all share the bathrooms."

"I don't think we exactly fit either category."

She smiled provocatively. "No, but our boats were docked next to each other."

"Our boats docked next to each other?" He repeated, thinking her comment a rather tenuous description of their relationship. Still, he saw no reason he shouldn't accept.

Inside, she said, "Start cleaning your hair, when I finish showering I'll help."

He caught brief glimpse in the mirror above the sink of her shapely behind as she stepped into the shower.

"How about joining me in a bottle of wine later?" he shouted, as he lathered his hair.

Above the noise of the shower, she replied, "I'd like that."

"We'll pick up a bottle from my boat," he called, "I suggest we go to yours, since mine is still a mess."

He was still working on his hair when she came out of the shower. After slipping into her robe, she told him she'd take over. "I don't want to go dancing tomorrow night with a guy with painted hair."

He was aroused by her nearness as she vigorously washed out the remaining paint. Finally, she pushed his head down and rinsed.

"That's much better." She turned off the faucet. "See you outside."

Cody removed his bathrobe and slipped into the shower. He soaped his hair to get any remaining turpentine out and any remnants of paint, then finished his shower.

Outside, he found Jeanette waiting.

"You look a lot better," she remarked.

"Thanks to you."

They walked down the marina as she continued to tussle her hair. On the way, Cody stopped to get a couple bottles of wine from the Columbia. As they climbed aboard Jeanette's boat, both in their bathrobes, Cody noticed Tony Pazzio watching them from his sloop at the far end of the dock.

When they settled in the cockpit, Cody opened the wine and said, "There's a tropical depression east of the Windward Islands. I wish you'd consider waiting until the hurricane season is over before you leave."

She smiled. "I've been thinking. Maybe I'll stick around awhile longer."

He glanced at the stars scattered through the rigging, and then out across the wetlands teeming with invisible nocturnal life.

"A penny for your thoughts," she said, and then added, "My mother used to say that."

He laughed softly. "Most people see the wetlands as ugly, nothing but grass and mud, and they hate the foul smell when the tide is out."

"Well they do stink, Professor."

"Not that bad. I like to listen to the wind blowing through the *Spartina* grass and the songs of the marsh birds. I like all of it, the bacteria, sea squirts, barnacles, clams, crabs, and the fish of course, and various insects and spiders, mollusks, and even the smell." He laughed softly. "There I go, boring you with all this."

"You're not boring me. But, like I said before, you ought to write a book."

He hadn't given it much thought when she first said that to him. Not a book, that is, but perhaps a series of articles for national magazines—magazines that would print environment-related pieces. He would reach a larger audience that playing articles in refereed journals or perhaps even writing a book. As he mulled it over, it occurred to him a book might come later, one that would grow out of series of articles.

He finished his wine and sat back, and glanced up at Polaris high above the city.

"It really bothers me that we think we can do what we want to the earth. We forget the planet is our home. We should treat it with respect. But no, we take what Earth has to offer, without regard to the consequences. Who gave us that right to mindlessly plunder the fossil fuels and our natural resources?"

He thought a moment then said, "I recall reading what Chief Seattle wrote about the earth not belonging to man, but that man belongs to the earth. He said all things are connected, and that man did not weave the web of life, but is merely a strand of it."

"I like that, Professor. I guess I never thought about things like that."

"Well, think about this: the universe somehow managed to bring forth intelligence, it follows that to produce this intelligence the universe itself has to be intelligent."

"That's pretty deep, Professor. And, you know, you're different. I never met anyone who talks like you."

'I'm concerned about the wetlands. They didn't come into existence to be destroyed for condos and shopping malls." He paused. "I'd like to talk to a lawyer about the possibility of a restraining order against the Gears."

"You may get a chance to."

He looked at here questioningly. She explained she'd asked Mitch and his girl friend to join them tomorrow night. "He dates around a little, and his current is a lawyer."

They talked awhile longer and then Jeanette said, "The darn no-seeums are getting to me," referring to the tiny, biting insects "We better get inside."

After they went below, Jeanette pulled a screen netting over the companionway.

Cody opened the other bottle of wine and refilled their glasses. As he handed Jeanette her glass he noticed her bathrobe had slipped open, revealing her thigh.

"I'm not much of a wine drinker, but I like this." She said, after taking a sip.

Cody raised his glass. "To the wetlands, to us, to Mitch, and to the whole marina."

"I'll drink to that," Jeanette said, smiling up at him. "Sit down, Professor." She patted the bunk.

"You better be careful," he said. "You're a very tempting woman."

She laughed softly. "Now I'm a temptress."

He took her glass and set it aside. Holding her hands, he drew her up and gently kissed her.

"That was nice, Professor," she said softly.

He kissed her again. Her mouth was soft and warm as he moved his lips over hers. Then, releasing her, he pealed her bathrobe off her shoulders.

"We better get more comfortable, Professor," she suggested.

She lay back on the bunk, as Cody removed his bathrobe and lay beside her. Gently, he caressed her breasts and then moved his hand slowly down now her body to between her thighs.

"Oh, God," she breathed,

He continued, enjoying her softness and her gentle exclamations. Then, as if by tacit agreement, he rolled onto her.

She was not at all like Karen, who was impetuous in her lovemaking, thinking only of her own pleasure. Jeanette moved her hips slowly in response to Cody, as though patiently enjoying every nuance of their lovemaking. She moaned softly, until, reaching her climax, she cried out.

Simultaneously, Cody exploded inside her. And for that moment nothing existed except the two of them united in violent ecstasy that swept them up in a wild torrent of desire and fulfillment.

Afterwards, lying together, she gently drew designs on his cheek with her fingertip.

Finally, he said, "I have a long day tomorrow. Mitch is putting me back in."

"Why not stay with me?"

"That is a very enticing suggestion."

He got up, sipped into his bathrobe and reached for the wine. "Let's finish this, first."

When they'd finished the bottle, she said, "Don't forget our date, Professor."

"Don't worry."

On the way up through the marina, he thought of Ben, whom he hadn't seen since their last encounter.

And, as if his thoughts could materialize the man, he heard Ben's voice.

"Hey, Matheson, I want to talk to you."

Cody turned to see Ben walking toward him.

"We got nothing to talk about," Cody replied, and resumed walking toward his boat. But fearful of Ben's intentions, he glanced around for anything that would equalize the disadvantage he had facing the more muscular man. Ben was now nearly up to him.

"You been on Jeanette's boat, you bastard," Ben growled, as Cody turned from him and untied his bathrobe belt, which he wrapped around his right hand.

"We're going to settle this for good," Ben threatened, as he brought up both fists and dropped into a crouch.

Cody turned and quickly thrust his left hand into the air. Ben was momentarily distracted, which gave Cody the advantage he needed. Cody brought his hand, wrapped with his belt, up from behind his right knee with all the strength he had.

His fist struck Ben's cheek.

Ben reeled backward and fell.

Cody was surprised the blow had knocked Ben down. He expected him to recover quickly and get up, but Ben lay on the ground for several moments with a look of surprise on his face.

Slowly, he got up on his hands and knees. He shook his head. Then he struggled to his feet, and without a word, went unsteadily back down through the marina.

Cody watched him disappear, and then continued toward his boat, glancing at the rows of boats resting on their iron stands like giant steel-legged dinosaurs of the night.

On board, as a precautionary move, Cody pulled up the ladder, and laid it on deck.

CHAPTER SEVEN

The old Chevy, with Cody driving, laid a wake of blue exhaust down Riberia Street in Saturday's fading light. He muttered that the guy who left it could have at least put in air conditioning.

"If you knew him you wouldn't say that," Jeanette said. "People called him 'nature boy.'"

Ten minutes later Cody turned into the Palms complex. and slowed to avoid hitting late golfers returning in their golf carts. He turned into a curving drive in front of the restaurant.

After letting Jeanette out, he found a parking spot, and then rejoined her in the foyer. Together they walked into the lounge. Cody glanced at the walls covered with murals of cavorting harlequins. The mirrored ceiling displayed an upside down world. The decor, along with the bar's polished brass foot rail, suggested something out of the early twenties.

He motioned to a table close to the dance floor, and when they were seated, a leggy waitress in a short skirt took their order, an Old Fashioned for him and a vodka and tonic for Jeanette. Cody saw the look of disapproval in Jeanette expression as he pulled his gaze from the shapely derriere retreating toward the bar.

The combo began playing a slow song. "How about it, Professor?"

"The moment of truth."

He found he hadn't forgotten what he learned when he and Karen had taken lessons.

"You're anything but rusty," Jeanette remarked, as they moved across the floor..

"You make it easy."

She smiled, her eyes sparkling. He told her she looked good in her wrap dress with its tiny floral design.

"Well, thank you, Professor."

He liked the feel of her close to him. The subtle smell of her hair and the spicy fragrance of her perfume made him recall last night when they made love. He pulled her closer against him as they moved to the music.

When the combo shifted to a fast number, Cody advised they sit it out. They found their drinks waiting.

"You've been here before?" he asked, curious how she knew about it.

"Once, with Marion."

He asked her what she planned to do after she finished photographing the islands.

She shrugged. "I guess I haven't thought that far ahead."

"Ever think about opening your own studio?"

"Not really."

"Will you go back to New Jersey?"

"I don't think so. There's nothing for me there."

Cody glanced around the room. Surprised, he saw the Pazzio's. "Look who's over there."

They sat off to one side at a table against the wall. As their eyes met, Maria smiled and Tony Pazzio nodded.

Jeanette waved. Cody remembered the night he saw Pazzio watching them.

"Must be a popular place," he commented, and added, "What do you know about them?"

"Dad assigned him to look after me?"

Cody shot her a questioning look. "He what?"

"I guess that needs an explanation," she said.

"Well, I think so."

"I should have told you before."

"You in some kind of danger?"

"I don't think so, but," she made an off-hand gesture, "Dad doesn't take any chances. Not in his line of work."

This would explain, Cody thought, why Pazzio had been watching them the other night, and, of course, his presence here tonight. Cody assumed Pazzio had informed Jeanette's father about him.

Jeanette changed the subject. "Is Mitch in danger of losing his marina?"

"It won't be easy for the Gears to force him out, and I'll sure as hell do everything I can to prevent it."

Cody had thought a lot about what the Gears were doing to the wetlands and their efforts to get hold of Mitch's marina. So far, he hadn't come up with any ideas of how to stop them.

He finished his drink and beckoned the waitress. "We should have done something to protect the wetlands long ago. It would have been easier then. We had the chance to adopt sound ecological practices and enact laws against destroying the land. But no, we decided to let the developers exploit nature."

"Is it too late?"

"I like to think it's never too late,"

She rattled the ice cubes in her glass. "I guess we're caught up in the course of our own events, you with your wetlands and me with my plans to capture the Caribbean on film."

"Like pawns of fate."

She smiled. "Maybe some things are meant to be."

"Maybe," he said, "our lives are written in some Akashian book of records. However, I don't like the idea of being a puppet on a string. However, I guess we make our own reality."

"Pass that by me again, Professor."

"It's out of quantum physics. The mind creates what we see out there."

"Have I created you?"

He laughed. "Have you ever thought what the world would look like if you could see all the wave lengths of light and hear the entire spectrum of sound? That would be one crazy world. Fortunately, we see only in a very narrow range. Which is all we need."

"Interesting, Professor."

"How about this: we can't discount the possibility that mind creates matter."

"Sometimes I don't know whether you're joking or serious."

"It has been demonstrated that the mind can affect things."

"Can I affect you with my thoughts?"

He smiled. "You don't need your thoughts."

"Oh? Would you care to explain?"

"I better not go into that."

When their drinks arrived, Cody continued. "But putting thought aside, the radical environmentalists and the Earth Firsters figured they could make changes by acting against violators of the environment. Perhaps that was the way to go. I thought it was better to preach my philosophy from the classroom or go through the courts. Now," he shrugged and looked across the lounge, "I don't know.

"I sometimes wonder if the environmentalists have gone too far, maybe lost contact with reality, especially when they came up with the idea of establishing a bucolic global village where everyone lives in communities, subsisting along with nature."

"Like turning us back to hunters and gatherers."

"Unmaking civilization." he said. "Nevertheless, they got people worked up over the desecration of the environment. I wouldn't say their efforts have been in vain."

"Shall we dance?" Jeanette suggested.

As they moved easily across the floor he again liked the way she smelled and the enticing softness of her body against him.

When the combo picked up on a fast number they went back to their table.

"They look like they're stricken with uncontrollable paroxysms," Cody remarked, watching the dancers gyrating across the small floor.

She laughed. "You have quite a sense of humor, Professor."

"I'm afraid where I took lessons they didn't teach that kind of dancing."

Once again, as the band changed to a slower number they got up and danced again. He felt more confident now and tried a couple different steps.

"You're really good, Professor."

"Thanks, but what you're experiencing is the limit of my expertise."

Then he saw her.

She was watching them. For a moment, their eyes locked. Even in the dim light he could see she was beautiful.

As they returned to their table, Jeanette said, "You know who that is?"

"Who?"

"The one you've been eyeing."

"I wasn't eyeing anyone."

"Well, you certainly fooled me."

"Okay. Who is she?"

"She's Ed Gear's wife. And the guy with her is our loveable Ed Gear."

Surprised, Cody looked more closely at the two before he sat down. Gear had ample light hair and sharp features. Actually, not a bad looking guy. His wife had blonde hair and, Cody expected, a good figure to go with her beauty.

"You can stop staring," Jeanette advised.

Cody turned as Mitch spoke. "Well, finally we made it." A very attractive woman was with him.

It has been a long time since Cody had seen Mitch shaved and his hair combed. He looked transformed in his blue sport shirt and light gray slacks, almost unrecognizable from the character who ran the marina and scratched his crotch. He half expected him to reach down to perform the ritual before being seated. Cody started to rise, but the woman motioned him down. "Please, no."

Cody liked the sound of her voice. It was soft, melodic. He noticed her deep blue eyes, dark hair, full mouth, and a good figure beneath her light one-piece print dress.

Mitch made the introductions. Her name was Shawna Gregory.

"Mitch told me about you." She said looking at Cody.

Mitch beckoned for the waitress and ordered another round.

"Don't believe all of it," Cody remarked lamely, feeling he could do better than that.

"Well," Mitch suddenly remarked, having noticed the Gears, "look who we have over there."

"The professor has already conducted an appraisal," Jeanette said.

Cody caught the iciness in her voice.

"Professor?" Shawna asked with an expression of interest.

Cody explained he had been with Virginia Tech.

Before she could ask what he taught, Jeanette said, "He has a PhD in marine ecology." She linked her arm through Cody's.

"Shawna," Mitch said, "is with the attorney general's office."

Cody wondered about Shawna and Mitch's relationship.

"I assume," Cody addressed Shawna, "that Mitch has legal title to the marina, so that the Gears don't have a means of acquiring it, at least not legally."

He wondered if perhaps he shouldn't have brought up the subject at this time.

Her expression showed she didn't think it inappropriate. "Everything's by the books. Mitch is running a good business; he meets his obligations and has clear title to the property. The only way they can get the marina is outside the law"

"Meaning?"

"I'm not suggesting anything, but I don't trust the Gears very far."

"What are the chances of getting an injunction against them?" Cody asked. This had been on his mind since he'd arrived.

"Can you be more specific?"

"The construction of the condos will destroy the environment. Isn't that grounds for legal action?"

Their drinks arrived and Mitch raised his glass. "To us," They all joined him in the toast.

"I wish there was," Shawna said, putting her glass down, "but the law says you have to go after the permit granting agencies, such as the State Department of Environmental Regulations, the fisheries and wildlife people, and the corps of engineers.

"So you can't go after Gear directly?"

"Well, only if we catch him breaking the law. And then there are the judges."

"Judges?" Cody questioned.

"Two of the judges on the fourth circuit court are direct opposites. Judge Henry Klein is an environmentalist. He would be sympathetic to an injunction, if that were possible. The other, Crutch Henderson, is in tight with Gear and his crowd. It would be hard to pass anything against the Gears in his court."

"It seems the Gears are playing a strong hand," Cody remarked.

Shawna's expression told him she wasn't happy about it. "Gear owns the land and he has the okay to clear it and build his condos."

Mitch broke in. "Hey, let's not talk about them. It'll spoil my evening." He turned to Jeanette, "how about a dance?"

When they were alone, Shawna said, "It was nice of you to come down to help Mitch. He told me about you and that you ran into a storm."

Cody referred to the broken shroud. "That's how I got this." He pointed to his forehead. The wound was fairly well healed, so was the bruise on his cheek from the fight with Ben.

"Care to dance," Cody asked.

She smiled, rising immediately. As they picked up the tempo, she said, "You're a good dancer, Cody, or should I call you 'Professor'?"

"Jeanette goes for nicknames. Whichever you like."

"I like Cody."

He saw Jeanette watching them. Mitch seemed oblivious to everything as he danced with his eyes half closed, a somnambulist ambling through erotic fantasies.

"I gather there's no Mrs. Cody."

"Not now."

He maneuvered Shawna around another couple and told her he was divorced.

"I want to help Mitch," he said, changing the subject. "He's a hell of a nice guy, and he cares about the environment like I do. Also I hate what's being done to the country's wetlands. I've done most of my fighting in academia. I had hoped that by changing young minds I could help save the environment. The problem is students don't give a damn. We live in a fast food society with a channel-surfing mentality."

She laughed, leaning back to look up at him. "You probably have learned that the Gears carry a lot of weight in this town. Fighting them means being up against the city's good old boys, the mayor, the city commission, not to mention a state senator." She paused. "My office stands ready to help in any way possible—legally, of course."

"I appreciate that. Is Mitch a client?"

She moved closer.

"Mitch and I are old friends. I've known him and his late wife for some time." Then she surprised him by adding, "She's crazy about you."

It took him a moment to catch her meaning. "Jeanette lives at the marina."

"Then you've known her only since you arrived?"

He told her that Jeanette doctored his forehead and that she was a photographer, planning to do a series about the Caribbean.

Then he asked, "Where did you go to law school?"

"Gainesville. The University of Florida Law School."

The music ended too soon. Shawna squeezed his hand when they got back to their table. "I enjoyed that."

Mitch returned with Jeanette. "Excuse me folks, got to make a trip."

"I'll join you," Cody rose.

On the way to the men's room they passed the Gear's table. The president of Condo Developers raised his hand. "Mitch, I want to talk with you. How about meeting me for lunch next week?" His eyes shifted from Mitch to Cody. "I don't think we've met."

"Dr. Cody Matheson," Mitch said.

Gear's eyes suggested nothing as he extended his hand. He nodded at his wife. "My wife, Erica."

She beamed Cody a warm smile. "You're not from around here?"

Cody told her no, he had been with Virginia Tech before coming to St. Augustine.

"You taught?" Erica asked.

"Marine science."

Mitch returned to Gear's earlier suggestion. "You know damned well there's no reason for having lunch."

"Perhaps," Gear said, "but no point in ignoring the situation, is there? Tell you what, I'll have my secretary set up a date."

Mitch's face clouded. "Don't bother."

Gear's eyes grew hard, but he forced a smile. "I'm offering you a good price for that place. You better think it over."

"Yeah," Mitch replied and abruptly left.

Erica Gear smiled up at Cody. "It has been a pleasure," she said. "Perhaps we'll run into each other again."

"Perhaps."

Ed Gear nodded without comment.

Cody joined Mitch in the men's room, and from the adjoining urinal, Mitch said, "That bastard won't give up."

"I saw that."

"Let me give you some advice. Be careful of that bitch. I saw the way she looked at you."

"Damn good looking woman."

Mitch muttered something unintelligible as he zipped his fly. "Any more trouble from Ben?"

Cody said no. He didn't mention the latest incident.

"You let me know. I'd just as soon get him out of the marina."

Back at the table, conversation went from boating to the economy and to the government's ineffective track record for preserving the environment.

"I think we have to face it," Cody said. "The governments, both state and federal, are pandering to big interests, such as timber and mining, and the government's stupid attitude toward the environment is creating havoc that will take years to correct. It's the same old thing. If you have money and pull, you can do anything."

He downed his drink, wishing he hadn't gotten off on the environment again. He motioned to the waitress.

After their drinks arrived, he asked Shawna to dance again. Jeanette watched them. When they returned to the table, Jeanette threw him a searing look.

At eleven, Mitch glanced at Shawna and said he had a lot of work tomorrow. "No rest for the wicked," he remarked with a grin.

Shawna told Jeanette she was pleased to have met her and hoped to see her again. She smiled warmly at Cody. "Thanks for the dances."

When they left, Jeanette settled a cold eye on Cody. "What went on out there on the dance floor, Professor?"

"She raped me."

"I saw how close you danced and the way she looked up at you."

He laughed. "Oh, come on, Jeanette."

She sipped her drink, regarding him malevolently over the rim of her glass. "Why don't you ask me for a dance and we'll dance close like you two did."

They danced and he held her close. The feeling of her body against him took his mind off the Gears and Shawna.

Back at their table, Jeanette remarked, "I've enjoyed the evening."

"So have I."

She emptied her glass. "I'm ready."

As they made their way toward the entrance, Cody saw that the Pazzios had left.

The temperature was cooler as they drove back to the marina. After parking, they walked to their boats. Mitch had put the Columbia back in the water next to Jeanette's boat. Cody wondered if he should ask her aboard.

"It's not late. Want to have a drink?"

"Sounds enticing, but I'm tired, Professor."

He was a little disappointed, but it had been a full day for him.

As he undressed, he had a little trouble putting Jeanette out of his mind. When he lay down, however, he began thinking about the Gears. Despite all he'd heard regarding their power in the community, there had to be a way to stop them from ruining the wetlands. They seemed to have everything on their side, the law, authority, a good old boy network, even the police and sheriff. It appeared they'd thought of everything. Before falling off, he recalled dancing with Shawna.

CHAPTER EIGHT

The sound of pounding awakened Cody from a nightmare. He dreamed an unruly, jeering crowd prevented him from entering the wetlands where the condos were under construction. They shouted obscenities as he struggled to get through, and their hands clawed at him liked tangled vines. The more he struggled, the more he became enmeshed. Up ahead he saw the Gears on a raised platform laughing at him.

Cody knew it was daytime from the light in the cabin, but he could only guess the hour. He recalled the previous evening at the Palms.

Again the pounding on the hull of his boat.

"Hey, Professor!"

With a towel wrapped around him he looked out the companionway. His head hurt. He'd drank too much. Jeanette stood on the dock, her face etched with a worried expression.

"Someone broke in last night."

"Broke in? Where?"

"Mitch's office. Jugs told me. Mitch called the police."

Why, he wondered, would anyone break into the marina's office? It was unlikely, those living here at the marina would. Perhaps teenagers, thinking they would find something valuable.

As he quickly dressed, he thought about the incidents Rocky and Mitch had mentioned.

Fastening his belt, Cody dropped down onto the dock. "What's this all about?"

"Jugs told me that Mitch came to work and found the office ransacked."

They hurried up through the boat yard, passed the boats on their metal stands, to where a group of live-aboards stood outside the office. Jugs swayed like an agitated elephant. Tony Pazzio, cast cold eyes on Cody and nodded. Marion Talbert, her auburn hair uncombed, stood beside Larry Mackintosh, who was babbling as usual. Jules Rodreques and Water Rat were talking together. Water Rat held an open can of beer.

Cody was about to ask if anyone knew anything, when a police car swung into the yard, its lights flashing. As it skidded to a halt,. it sent up a cloud of dust, A short, over-weight officer climbed out of the driver's side, hefted his gun belt, and looked around.

"What's going on here?"

Jugs motioned toward the office. The other officer, a young woman, turned up her pug nose in disdain, as she also got out and stared at the group with a look of authority.

"You folks stand back," she ordered, her dark, piercing eyes flashing. "Who knows what happened?"

Jugs spoke up. The office got broken into.

"Oh? By whom?"

It was, Cody thought, a stupid question,

"Anyone know who did it? the female officer persisted, glaring at everyone."

"How should we know?" Cody answered, wondering why the officers didn't go in and check it out.

The lady cop looked at him scornfully. "You the elected spokesperson around here?"

"You asked a question."

"And I expect an answer. You know anything about this?" She leveled her eyes on Cody.

"I was told someone broke into the office."

"Then I think we better find out about this."

She and the other officer entered the office.

Cody and Jeanette followed and stood in the doorway, when the officers had entered. Furniture had been overturned, supplies scattered everywhere.

"Christ," the overweight officer remarked, "what a mess." The two cops stepped gingerly around the debris. " When'd this happen?" the female officer asked.

"This is what I found when I came in this morning." Mitch explained, looking like he didn't believe it.

"They pulled out my files. Looks like they took my records."

The female cop studied him. "They?"

Mitch shook his head. "Whoever did this."

"But you said 'they.' You have some people in mind?"

"No," he replied.

He sure in hell does, Cody thought.

"Well," the other officer said, "Whoever it was did a thorough job."

Brilliant, Cody reflected.

The two began to conduct what passed for an investigation.

"No telling who could have done this," the male officer remarked finally. "Probably used gloves so there won't be any prints."

"Wouldn't it be advisable to dust for them?" Cody asked.

The female officer spun around. "What's your name and what is your business here?" When Cody told her, she retorted hotly, "Why don't you leave the investigation to us. We know what we're doing." She confronted Mitch. "Your not having any idea who might have done this, doesn't help."

Both cops checked around a few more minutes. They took no notes.

"You got a list of everyone in the marina?" the male officer asked. Mitch said he had, but couldn't produce it until he got the place cleaned up, and only if it hadn't been taken. "Okay, fine, you give us a list when possible."

It was obvious to Cody the police weren't interested in wasting any more time.

"I don't think anyone at the marina did this," Cody ventured, knowing how that would be received.

The fat male officer looked at him. "Why do you say that?"

"Well, for one thing, why would anyone at the marina trash the office. They'd have nothing to gain. This was done from the outside."

"And I suppose," the other said sarcastically, "you have a suspect?"

"I might."

"Oh?" The fat cop fixed his small eyes on him. "Care to say who?"

Cody knew it was too late to back out. "Someone who has a beef with Mitch, would be my guess."

The female officer regarded him closely. "And who might that be?"

"Well, there is someone who wants this marina. Wants it bad. I think I'd check that out."

"You think we should check that out," she intoned. "Like you're telling us how to run our investigation?"

He'd like to wring her neck. "I didn't say that."

"Are you going to name names?" the fat officer asked.

Damn, Cody thought, he should have kept his mouth shut. "The developer who's putting up condos across the way wants this marina. The Gears."

"You think they trashed the office?" the woman demanded.

"I think it is a lead that you should check out."

"So you're accusing Ed Gear of the break-in and damage to the office?" the fat one asked.

"I wouldn't put it that strong."

"Then how would you put it?" the woman asked.

"Look, all I'm saying is that from what I hear Gear wants this marina as part of his condo complex. Wouldn't that suggest a motive for what's happened?" Cody knew, of course, Gear wouldn't do this himself; he would have hired someone.

Both officers eyed him closely. "Maybe we ought to take him down to the office," the woman suggested.

The fat cop studied Cody. "I think it would be a good idea that you be available in case we want to ask you some questions."

They wrote down his name.

The fat cop glanced at Mitch. "Sorry about this, Mitch." The two went outside and climbed in the patrol car.

With the look of a condemned man, Mitch watched them leave. The others crowded around the entrance.

"You know goddamn well who did this," Cody said.

Mitch nodded. "If that son of a bitch thinks this will change my mind, he's a dumb shit."

"You don't have duplicate records?"

Mitch looked dazed. "Christ, this could put a guy out of business."

"You've checked to make sure all the records are gone?" Jeanette asked.

Mitch scratched his crotch. "Seems like it. But I'll look some more."

Both Cody and Jeanette, along with several others, offered to help.

"I appreciate that," Mitch said, "but looks like I have to do this myself. I know what I'm looking for." Then he added, "Fortunately, a buddy of mine put some of my files on a computer disk. I hope the hell he still has them."

Cody and Jeanette walked back to their boats.

"It's terrible," Jeanette, remarked, "that the Gears would do such a thing."

"I realize now I wouldn't put anything past them." Cody rubbed the stubble on his face and knew damn well the police would not follow up on the break-in. Probably they'd file a brief report and forget about it.

"We could have breakfast at the cafe," Cody said, "but why don't I put something together?"

"I'd like that, Professor."

Aboard the Columbia, Cody turned on the burner of the small stove and got eggs out of the 'fridge.' He saw Gear in a different light, not that he'd previously regarded him in any favorable way. But before, he had been without a defining personality, an amorphous enemy. Now he saw him as totally ruthless, motivated by his own goals regardless of the harm to others.

"That took nerve, Professor, pointing a finger at the Gears." Jeanette said. "I imagine the break-in was done while we were at the Palms."

"Or later, when everyone was asleep."

Cody broke several eggs into a bowl. By being at the Palms, Gear has an alibi.

Jeanette watched him put diced potatoes in another pan and sprinkle them with condiments.

"It would be hard to prove he was behind it," Cody remarked as he put two slices of bread in the toaster. "Then again, the way he's got things sewed up in this town, I doubt anything would be done about it."

"I hope Mitch can retrieve his records."

Cody's began thinking about an idea that had occurred to him while he had viewed the ransacked office. By the time he put the food on the small cabin table, he had it worked out.

"You got something on your mind, Professor," Jeanette said, spearing a piece of potato.

He sipped his coffee. "We can't prove Gear instigated the break-in, I mean, we just can't accuse him outright, but perhaps we have an alternative." He pushed some scrambled egg onto his fork with a knife.

"Can you be more specific, Professor?"

"Suppose we take a different approach."

"Okay, what?"

"The area where Gear's putting up the condos is a natural habitat. The city thought it was nothing but a damned swamp and was glad to get rid of it."

"We already know that, and we know Gear doesn't give a damn about the land or the wildlife."

"But that's just it."

She stared at him, her fork poised just short of her lips.

CHAPTER NINE

With breakfast over, Cody was impatient to put his plan in action. Jeanette volunteered to do the dishes.

"You going to explain what you're up to, Professor?" She asked, clearing the table.

He got a pad and pencil from the shelf above his bunk. "When I get back, if it works out."

"You can at least tell me where you're going."

"When I get back."

"At times you can be very mysterious, Professor."

"Well, not intentionally," he replied, as he put an arm around her.

As he dropped down onto the dock, Cody watched a pileated woodpecker dart into the protection of an oak. A long-billed dowager stalked along the shore and several cormorants sunned themselves atop the rotting posts of forgotten piers. A pair of kingfishers swept by, and a small flock of wood ducks circled beneath a sky of thin wisps of cirrus. He felt the sea breeze against his face. The day was already hot and humid.

Cody walked up through the marina and crossed the street that ran between it and the development site. As he entered the site, he paused and ran his hand through his hair. It was Sunday, there would be no workers in the area, but he glanced around to make sure. The air was heavy with the pungent odor of the wetlands now exposed at low tide. He walking through the area, glancing at the stands of cypress, pine, palmetto and water oaks interspersed the wetlands. Many trees on firmer ground had been cut to make a road.

Although he didn't know how the final complex would be laid out, he was aware Gear would have to rework the landscape drastically. In addition, a lot of fill would be needed for the building foundations and parking lots.

Little red flags indicated where foundations would be placed. Paint splotches marked trees to be cut. A hundred yards into the site, he came to the end of the road. Beyond it through the trees and undergrowth lay the Intracoastal Waterway. Several narrow channels allowed the water to ebb and flow into the area.

Cody thought of the centuries that the advance and retreat of the tide had formed this wild and beautiful wetland. A vast battery of nature's engines fueled by carbon dioxide, water, mineral nutrients from the soil, and energy from the sun produced tons of wood each season. During one season's growth, an acre of trees drew up 4,000 tons of water from the soil, and the trees held the earth from washing away in the tropical rains.

He squatted and looked closer at the edge of a channel where small blue crabs scurried through the mud. Looking closely, Cody spotted lug worms. Further inspection revealed razor clams, burrowing shrimp, and sea squirts, and barnacles. Small fish darted away as his shadow fell across the water.

Cody made notations on his pad.

He also checked the composition of the wetland soil. Most of it was held in place by the *Spartina* grass and the roots of trees. As he scooped up a handful of dark mud and rubbed it between his fingers, he knew that without the vegetation the soil would wash away in heavy rains or tidal flooding.

He thought about the marsh of his youth. Like that one, this was a habitat where specialized life forms had developed in strange and fascinating symbiotic relationships.

Algae lived in the black ooze that, along with plant life, was the major food producer. Cody saw grasshoppers, cinch bugs, beetles, and flies.

Bacteria feasted on the dead grasses and turned them into a detritus-algae soup that became in turn food for fiddler crabs, mullet, mussels, oysters, and insects. Wrens lived here with sparrows, wallets, and rails. Larger birds, such as ibis and gulls, sought out the crabs, snails, and worms, while marsh hawks cruised looking for mice. Each day the tides washed part of the marsh's production into the creeks and bays where porpoises, fish, and shrimp consumed it.

He noted the herons and snowy egrets that prowled the grasses. Hordes of mud crabs emerged from their burrows, the males waving their single large claws in a bizarre courtship. Mud snails left wandering trails in the ooze.

Anger formed a hard knot in Cody's gut. He could not let Gear destroy all this. Yet he had to admit he didn't have a plan of action, other than what he now had in mind.

When he had taken sufficient notes, he headed back toward the entrance. He was surprised when a large man in a white hard hat appeared from around the curve in the road into the site.

"Need help, Buddy?" He looked suspiciously at Cody, as he approached.

"No, why should I?"

"This is private land. What are you doing here?"

Cody feigned exasperation. "You should have been told; I'm from the survey company. I have to check the boundaries."

He made a sweep with his arm. What was this guy doing here on Sunday?

The man regarded him narrowly, glancing at the pad in Cody's hand. "No one said nothin' to me. I was told to look at a problem with one of the cats."

"Well," Cody replied, "I got work to do. If you don't mind, I'll get on with it."

The fellow studied him a moment. "You should be wearing a hard hat." He was trying to exert his authority.

"If I had one, I'd be wearing it," Cody retorted.

The man grunted, threw him a final look, and moved on. Cody walked unhurriedly out of the site and back to the marina. Before returning to his boat, he stopped at the marina office. Mitch looked up from his work of cleaning up the place. Those helping him used the occasion to pause in their work to rub their backs.

"Can I borrow your typewriter?" Cody asked.

Mitch scratched his crotch. "You're welcome to it. Looks like I won't be using it for a while."

Aboard the Columbia, he found Jeanette had cleaned up and left. He set the typewriter on the small folding table and laid out his notes. The next couple hours he typed. When finished, he read over what he'd written. It would be a good idea to get another opinion. He went over to Jeanette's boat.

"How about being the devil's advocate?" He handed her the sheets of typed paper.

"What is this?"

"Read it."

As she did, Cody watched her. She nodded occasionally and once or twice raised her eyebrows.

Finished, she looked up. "I like it, Professor." She glanced down at the sheets before handing them back. "So this is what you've been up to. What do you plan to do with it?"

"Get it printed."

"Where?"

"I hope in the local paper."

"You won't make points with Gear and his buddies with this. But," she smiled, her eyes sparkling, "I doubt if that was your intention."

————-

The letters on the glass door read: Managing Editor, St. Augustine, Record. As Cody entered, he noticed the afternoon sun slanting through the dusty window behind the editor's desk.

After motioning Cody to a chair, Mark Wilson sat back, his brown eyes leveled on his visitor. "You phoned you had something I would be interested in."

Cody judged him to be in his forties. He had receding hair and a square face.

"I put this together." Cody handed him the article.

Wilson started reading, Cody explained, "Where the condos are going up is a natural habitat, though not officially designated as one," Cody said.

"I'm familiar with the area." the editor replied without looking up.

He read fast and after finishing, tossed the pages on his desk with a finality that Cody interpreted as not good.

"Looks like you covered it thoroughly enough. I trust there's no bull-shit here."

"There isn't."

"You know Ed Gear?"

"I've met him."

Wilson glanced back down at Cody's report. "I can see the shit flying."

Cody did not comment.

"Some people don't like what he's doing. Some do. You have to understand the newspaper business." He pulled a cigarette out of its pack and offered one to Cody, who shook his head.

After Wilson clicked his lighter shut and exhaled a cloud of smoke, he said, "Gear has a lot of influence around here and he's not the type to take something like this lying down." He tapped Cody's material with the backs of his fingers.

"You see, we have responsibilities to our readers, to all readers." He puffed thoughtfully. "That project is a big one, as I'm sure you realize, and important to St. Augustine. It means money, jobs." He waved a per-functory hand to dispel any comments Cody was about to interject.

Cody interjected anyway. "You're forgetting the environment."

"I'm forgetting nothing. I'm an environmentalist myself." Flicking an ash, he went on. "Not as rabid as some of those ecotage freaks. However, you can't be in this line of work and not know what's going on."

Taking a long drag, he studied Cody. In the silence, Cody decided he'd tell Wilson what he thought about his claim to being an environmentalist, and what he thought about his paper.

Before he could, Wilson picked up the story and glanced through it again. His expression conveyed nothing. The man, Cody thought, would make a good poker player.

Cody saw no point in waiting for more from Wilson. He got up.

"You got to understand I can't use this," Wilson said, looking up at Cody. He ground out his cigarette and reached for the pack and extracted another.

"So much for your interest in the environment," Cody retorted.

"Now, just a minute," the editor said. "I can't use this as it's written. We got our own way of doing things."

"I'll put one of my reporters on this." He glanced down at the material. "You've given us some good basic information. We'll build on that. But from our point of view.

"We'll refer to you as the authority. We'll quote you throughout." He rested his dark eyes on Cody. "That okay?"

Surprised, Cody replied, "That's fine."

"I want my writer to go out to where they're putting up those condos. I also want him to talk with you. I want him to get quotes from you.

"Of course we'll want some quotes from Gear, but we'll handle that." He smiled craftily. "Gear won't be aware of what kind of piece this is going to be. If he did, he'd throw my reporter off the place, and that would sure cause all hell to break loose."

Cody didn't care how they handled the story, as long as they presented a case for the wetlands.

Wilson got up and came around his desk. He took Cody's arm and walked him to the door. "Appreciate you bringing this to my attention.

You know, we did a piece on that project two or three months ago. At the time, it looked like a good thing, like I said, jobs, money, tourists, and all."

Cody felt like he'd hit a home run and did not conceal his pleasure as he shook the editor's hand.

"I hope you approve of how we handle it, Doc."

"I hope so. I thought for a moment…"

"Yeah," Wilson interrupted, smiling, "You didn't wait for me to finish."

Cody grinned as he walked to the old Chevy. He hoped that when the story came out, other papers would pick up on it. If they didn't, he'd send them copies. Maybe send copies to the radio and TV stations.

He felt good. At last he was doing something besides preaching to a bunch of apathetic students.

Cody indulged himself with thoughts of Gear's reaction. He certainly wouldn't like it, as Wilson suggested, and there was no question that Gear wouldn't take this kind of publicity without doing something about it.

CHAPTER TEN

Three days later, the front-page story in the St. Augustine Record read: ECOLOGIST PREDICTS DISASTER The subhead added: *Dr. Matheson predicts Planned development will destroy environment.*

The by-line indicated Jacob Reading was the author, the reporter Wilson had assigned to the story.

Cody sat in the cockpit of his Columbia and read the story in the morning paper.

The same moment, Erica Gear sat across the breakfast table from her husband. She had glanced at the article as she brought it in from the driveway: it caught her by surprise. She thought Cody an attractive man, one, perhaps, who could be manipulated. However, she had apparently misread him, which irritated her. She didn't like surprises from men. It was her habit of providing the surprises.

"I'll be a sonofa-goddamned-bitch," Ed Gear exclaimed.

He read aloud, "'According to Dr. Cody Matheson, an ecologist, the planned construction of the new condo development south of St. Augustine will produce an environmental disaster and destroy a natural habitat.'"

"Jesus Christ," Gear shouted. If he'd known what that sonofabitch reporter was up to, he'd never have allowed him on the site.

He continued: "'The project, which has encountered some opposition by environmentalists, is being built by Condo Developer, Inc. of St. Augustine. "'Edward Gear, the company's president, stated about one thousand units are to be built.

"'The project,' says Gear, "is to be unique. There'll be nothing like it. Part of the condo complex will consist of a natural wildlife sanctuary.'"

"The bastard got that right," Gear interjected.

"'But,' says Matheson, "the idea of enhancing a vast interior marsh is ludicrous.'"

"Bullshit."

"'The construction of the condominiums will cause the devastation of the natural forest land and cause irreparable damage to the area's ecology. The area covers a large part of the peninsular south of the city, extending to the Matanzas River.

"'According to Matheson,'" Gear continued, "'the destruction of the woodland will create a dangerous increase in soil erosion.

"'Another problem arises from run-off from the streets and parking lots as oils, tars, and other pollutants find their way into the ecosystem. Furthermore, pesticides and herbicides used on the grounds will add to the pollution problem. This does not include the deleterious effects on the ground water.

"'One major concern is the loss of wildlife.

"'The area is the home of the snowy egret, marsh wren, and the bald eagle, among other species. The planned construction will drive most wildlife away.

"'With the loss of the trees, Dr. Matheson estimates that the danger of flooding from heavy rains associated with thunderstorms and tropical storms will increases several fold.'"

Gear threw the paper down. "We got to do something about this."

"Any ideas?"

"Get rid of this sonofabitch."

"Be careful," Erica cautioned.

"Goddamnit."

"Well, my darling," Erica said, "this has to be handled correctly. We can't take any chances." She paused to sip her coffee.

Ed Gear felt a cold chill. He knew how the others involved with the project would react. The gutless wonders would be scared. They'd wonder how they could cover their asses.

Gear was right. The story left the watchdogs of the environment in shock. Burt Higgins, head of the Audubon Society threw animosity aside to call Ken Stoaks of the Sierra Club.

"I got my secretary preparing a release," Higgins explained, his three-hundred-pound body sweating profusely.

"Jesus, I wasn't prepared for something like this," Stoaks whined.

Higgins wheezed into the phone. "I suggest you get something out, too. We got to protect ourselves. It looks like we haven't been doing our jobs."

"Who the hell is this Matheson?" Stoaks asked.

"You know as much as I do. I'll make some calls and let you know. Point is this guy's caught us off guard."

"Damn. I'll get back to you." He hung up and eased his huge bulk back in his chair.

Their press releases would voice concern over the newspaper story, but with typical bafflegab, they'd confused the issue so no one would understand their positions.

Another person who had read the newspaper account with interest was Russ Gannet, nineteenth district representative. He didn't know Dr. Matheson, but he was impressed with what the man said.

Much less impressed was Mayor Douglas Orlando. Although he half expected the project would experience problems from environmentalists, he wasn't prepared for the indictment splashed across the front page of the Record.

"Where'd this guy come from?" he said aloud. His maid, an attractive, middle-aged black, whose duties, it was rumored, extended beyond housekeeping and cooking, shook her head as she poured more coffee. Angrily, he slammed the newspaper down, knocking over his cup. How would this make him look?

City Attorney Kent Daniels, suffering a terrible hangover, arrived late at his office. "Have you read this?" he asked his secretary.

City manager Brad McCarver groaned as the phone rang. He rolled over and grabbed it off the nightstand. The radio alarm clock went off. "Christ," he muttered angrily. Like Kent Daniels, he wished he'd taken it easy on his drinking the previous evening. His head ached; his stomach churned.

"You seen the paper?" The strident voice of Commissioner Wally Furgeson added to his frayed nerves.

"No, Wally," McCarver responded irritably,

When he hung up, he went out in his bathrobe and slippers and got the paper off the driveway. After scanning the article on the condo project, he took several aspirin. Like others, he had put a considerable amount of money into Gear's project, and having overextended his finances, couldn't afford to have anything go wrong.

While those associated with the condominium project were scrambling to decide what action was appropriate, some St. Augustine residents were amused. They regarded the local politicians as good old boys, concerned more with their own welfare than their constituents'. Like many people, however, these citizens did little more than complain, venting protestations in private. As for the city commissioners many people referred to them as "country bumpkins."

While most Floridians were fed up with Washington's cavalier attitude toward ecology, those who cared hadn't liked what Gear was doing, but had no effective way to stop him. They had not formed a phalanx of opposition, and cocktail affirmations seldom produce effective action.

The afternoon of the day following the appearance of the article in the Record, Mayor Douglas Orlando, City Attorney Kent Daniels, City Manager Brad McCarver, and City Commissioner Wally Furgeson met in the mayor's office with Ed Gear. It was Gear's idea.

Ed Gear and Erica arrived before the others, and when the rest showed up Gear told them about his phone call to Bill Gooney, head of the corps of Engineers. He chuckled, needing something to ease his tension, because he was still seething over the Record, article.

"I asked him about Andy," Gear related. "Remember the two of them were damn close during that conference we all attended dealing with land developments."

They remembered.

"He acted nonchalant. 'Andy?' he asked."

"You were pretty close," I told him.

"He didn't think that was important. So I told him it really wasn't, and reminded him of our need for the corps approval for some work we had to do. He didn't think there'd be any big problem. Of course, I had to be damned sure there wouldn't be. Then I asked him if he'd heard from Andy. Phone call? Letters? I asked him where they'd gone for dinner that one evening.

"He didn't see how that was relevant. And I said, 'well, when two guys go out for dinner…' I let it go at that.

"He said nothing for a moment, then told me I was a goddamn sonofabitch and hung up."

Gear now leaned forward with his elbows on the large desk. "I believe he got the message, gentlemen. So I don't expect any problems from the corps."

Gear then told them he'd heard from Blanchard.

"Don't know how he found out so quick. He was upset, and reminded me that he had too damned much money riding on the condo project to see it blow up and demanded I explain what in hell was going on in St. Augustine."

Ed Gear didn't like relaying Blanchard's conversation, but he figured telling the others how he'd handled the senator would make him look in control and confident.

"What'd you say?" Wally Furgeson asked.

"I told him we'd take care of it. We wouldn't let this Matheson upset our plans."

"He bought that?" Mayor Orlando queried.

"I think so."

"You know this Matheson?" McCarver asked.

Erica interrupted. "He lives on a boat at the Cormorant Marina."

Gear leaned back and crossed his legs. He was outwardly calm, but inwardly seething. "What's important is how we handle him."

"Looks like Mark threw us a curve. Really didn't expect it from him," Daniels lamented, referring to the Record editor..

"You can't trust newspaper editors." Furgeson put in.

"I think," Erica said, "we have to decide how to deal with this publicity. And quickly."

"You got any ideas?" Daniels asked.

"We've been discussing it," she replied.

"I've a couple alternatives," Gear responded. "Have any of you heard of SLAPP." He glanced at Kent, who, he knew, had.

"It's an acronym for Strategic Lawsuits Against Public Participation," Gear continued. He nodded at Daniels. "You explain it."

The others focused on Daniels. "It's a way of dissuading people from interfering in certain issues. I'll give you an example. The Pirini Land and Development Company, a Massachusetts firm, slapped a lawsuit on a Rick Sylverst, who opposed the company's building a large resort in the alpine town of Squaw Valley, California. The suit charged he conspired to overturn an agreement reached with other environmentalists.

"A $75 million lawsuit isn't something to sneeze at, gentlemen." Daniels lit a cigarette. "Most people just don't have the bucks to fight it."

"And there's another alternative," Gear said, his eyes hard.

Everyone looked at him.

CHAPTER ELEVEN

Cody looked up from the Columbia's cockpit, where he'd been typing an article using an old manual typewriter he borrowed from Mitch. It was about the desecration of the wetlands. The yard worker told him he had a phone call.

"Who is it?"

The guy shook his head. "Mitch just said to get you."

"Okay, I'll be right there."

Cody set the typewriter aside and walked up through the boat yard. When he entered the office, Mitch looked up from his desk and nodded toward the phone. He made an expression that conveyed he had no idea who it was. Cody noticed a new face in the office, a young woman, quite attractive, with dark hair and eyes.

"Yes?" Cody asked, when he picked up the phone

"Cody Matheson?" the man asked.

"That's right. Who is this?"

"Not important. You want to know about the break-in?"

"What do you know about it?"

"Well, it ain't me. I was told to set up a meeting."

"Who with?"

"Can't tell you over the phone."

"Why not. What's so secret about it?"

Mitch and the young woman glanced up at the tone of Cody's voice.

"Mitch owns the marina. Why not talk to him. Or the police."

"Look, I got my orders."

"You better tell me what this is all about."

"Like I said, I'm doing what I was told. You want to know about the break-in, I'll pick you up."

"Look, I'm not interested in playing games. Who wants to talk with me?"

Silence.

"I was told to set up a meeting."

"Like I said, who wants this meeting?"

"Hey, I'm just doing what I've been told. You interested?"

"Maybe, but you better fill in the details."

"There ain't no details."

"Then get them and call me back."

"I can't do that."

Cody was interested, but was getting exasperated. "Don't jerk me around."

"Hey, man, I ain't doing that, but I got my orders."

"From whom?"

"Jesus, you're making this difficult."

"All right, where we going?"

"Here in town. My contact has information." Then he added, "I'll pick you up at seven this evening. In front of the marina."

Cody glanced at Mitch and the woman. Both were still watching him intently.

Suspicious, but with his curiosity piqued, Cody told the man seven sharp.

"No problem."

The phone went dead.

"What the hell was that all about?"

"Someone wants to meet with me about the break-in."

"Why in hell didn't they talk to me?"

"That's what I asked."

"What's the deal?"

"They'll pick me up at seven."

"Who?"

"I haven't the slightest."

"Want me to come along?"

Cody tapped the cradled phone. "I'm not sure how that would go over."

"Think the guy's on the level?"

"No way of knowing unless I check it out"

"I don't like it. But you're right. We won't know what this is all about without going along with it."

"Maybe I should follow."

"I appreciate that, but let me check this out alone."

Reluctantly, Mitch agreed.

Promptly at seven, a blue Pontiac pulled up. A thin fellow with a small mustache rolled down his window. "Matheson?"

"You the guy who called?" Cody asked.

"That's right. Hop in."

He tossed his cigarette butt out the window.

Cody walked around to the passenger's side."Where we going?"

"The guy with the info doesn't want to be seen with you in public."

Five minutes later, the thin man swung into a parking space behind a St. George Street shop. As Cody followed him toward the rear door of a building, he felt uneasy. Still, it was possible whoever he was meeting did have information.

"Here he is," the skinny guy said, and roughly shoved Cody into a semi-dark room.

"What the hell…" Cody demanded.

It took Cody's eyes a moment to get accustomed to the darkness. Stacked boxes lined one wall and odds and ends of merchandise lay about the room.

Two men came toward him.

"We got a message for you," one said. They were big, muscular, dressed in jeans. Cody judged one to be over two-hundred and fifty pounds. The other, bald, smaller, but equally well built.

"What kind of message?" Cody asked, not at all liking the situation.

"This," the big one said. The blow caught Cody on his mouth, sending him reeling backward. The other closed in. As he did, Cody dropped to a crouch and threw a punch at the smaller man's mid-section. The guy grunted.

The other hit Cody with another powerful punch just below his right eye. Then the skinny guy who'd driven him here grabbed Cody's arms, pinning them to his sides. The big man moved in quickly, striking Cody repeatedly in the ribs and face. Darkness swirled around Cody's vision threatening unconsciousness. He collapsed on the floor.

"This is a warning," the other said, as he kicked Cody in the side. Pain shot through Cody's rib cage. He rolled, trying to avoid another kick.

The two men lifted him to his feet.

"You bastards," Cody shouted.

Calling on his remaining strength, Cody fired his fist into the big guy's nose.

Blood ran down his mouth.

Cody wrenched around and brought his foot up into the groin of the skinny guy who had pinned his arms. The man cried out and doubled over.

The door through which he'd entered was still open. Cody made a dash for it.

The blow from the two-by-four caught him between the shoulders and knocked him down. He rolled over and looked up at the smaller man.

"Thought you'd get out, eh? Well, you ain't getting off that easy."

"If you haven't gotten the point, shithead, it's to mind your own god-damned business."

"Yeah," the big one said, "like writing for the newspapers."

The two hauled him to his feet. His back throbbed where he had been struck and his ribs hurt like hell. He felt his face swelling. One eye must be swollen shut; he couldn't see out of it.

Summoning his remaining strength, Cody kicked the big man in the stomach. He doubled over. Grabbing the two-by-four, lying on the floor, Cody swung it at the other man, who ducked.

The skinny driver again caught him from behind, but Cody twisted around and fired his fist into the man's belly.

He figured now he could escape to the door and get out, but the big man tackled him and they fell to the floor.

Now the other was on his feet. Together they yanked him up and propelled him through the door and threw him onto the parking lot.

The driver came out, still hunched over.

"Jesus," he moaned.

"Shut up," one of the men said. "Drive him back to the marina."

Cody knew the fight was out of him. He struggled to stay conscious as they shoved him into the car. He vaguely recalled the ride. Blackness hovered around the edges of his world.

He felt the car come to an abrupt halt.

"End of the line."

The driver pulled him out and dropped him onto the ground, where he lay as the car spun around, throwing dust, before it hit the asphalt and sped down Riberia Street.

Cody lay with his face numb, covered with blood and dirt.

Slowly, painfully, he got to his feet and unsteadily made his way down through the marina. He saw no one until he reached the Brinsons' boat. Rocky was climbing down the ladder with a towel around his neck.

"Hello, Professor," Rocky said. But when he saw Cody's face, his expression changed. "What in hell happened?"

Cody attempted to describe the beating.

"Jesus Christ, you're a mess." He pounded on the hull of the boat, shouting for Millie to get the first aid kit and meet them at Cody's boat.

Rocky took Cody's arm. "You better fill me in,"

By the time they reached the Columbia, Cody had given Rocky the details. Millie joined them and swept Cody with a brief appraisal.

"For heaven's sake, Cody, you need attending to."

Jeanette had heard voices and came on deck. "Oh my god," she exclaimed, seeing Cody.

As they helped Cody below, Rocky told Jeanette what he knew.

Millie immediately took charge.

"Get the blood off his face, so we can see the damage." She told Jeanette to get some water. She and Rocky removed Cody's shirt.

With a cloth soaked in water, Millie carefully cleaned Cody's face. "This may hurt," she warned, "but that can't be helped."

Fortunately, not much feeling had returned. But his ribs and back ached.

"Should we call an ambulance?" Jeanette asked.

"Let's not jump the gun," Millie advised. She continued cleaning the blood from his face.

Cody's lips were split and swollen and his face bruised. He could not open one eye.

"I'm going to put antiseptic salve on your face," Millie said. As she applied it, Cody winced. When she finished, she checked the bruises on his rib cage. "You got worked over good."

She moved her fingers gently over his sides, but with enough pressure to check his condition. "You're lucky. I don't think you have any broken bones. I'm not going to tape you up. They don't do that anymore."

Cody managed a smile and wished he didn't have to breathe.

"You're not much to look at," Rocky said, good-naturedly, "but you'll live."

Jeanette gently kissed Cody's cheek. "You want anything, let me know. I'm bunking here tonight." Then she added, "Shouldn't we call Mitch?"

"There's nothing he can do for now," Rocky replied. "We'll brief him in the morning."

"I could use a drink," Cody said,

"That's a good sign," Millie remarked. They left as Jeanette mixed his drink.

"I'll be right back," she said, handing it to him.

She returned shortly with her nightclothes.

After making up the other bunk, she said, "I'll be right here, Professor. You want anything, just call."

He raised his glass. "Another."

"You'll be smashed, Professor."

"I'd welcome that."

When she made another, she said, "They can't get away with this. It was the Gears, Professor. It had to be."

It hurt to speak. "Yeah," he said.

"I think I should call the police."

"It would be a waste of time."

"Damnit," she swore. "Somehow we'll get them for this."

Those were the last words he heard as he slipped under.

CHAPTER TWELVE

The idea came to him after the beating. At first he was surprised a bit over the audaciousness of it, and wondered if it was spawned out of anger rather than clear thinking. Nevertheless, it seemed the only alternative against the Gears. True, it was not an approach he had seriously considered in the past, because he drew the line when it came to breaking the law.

That is why, as he and Jeanette watched the final floats of the St. Augustine Fourth of July parade, he was tempted to dismiss the plan in favor of fighting Gear through the courts. As parades go, it was a good one—local bands, fire engines, antique cars, horseback riders, the National Guard, a contingent from the navy, and a small group from the Marine Corps League. It reminded him of the parades of his youth in Marshfield.

Jeanette broke into his thoughts. "Let's get something to eat."

"What about the picnic?"

Mitch's annual Fourth of July bash was this afternoon. In addition to those living at the marina, Mitch usually invited a number of people from in town. Cody wondered if Shawna would be there.

"Oh, we'll pig-out at that," Jeanette remarked, "but in the meantime we need a little something, maybe a cup of coffee and a salad or sandwich, whatever."

They turned down a narrow side street to a small cafe.

"How's this, Professor?"

"Looks okay to me," Cody responded as he took her arm and assisted her up the steps.

As he did, he glanced up at the cumulus building to the west. There'd be rain before the day was out.

"How about over here?" Jeanette motioned to a table by the window, giving them a view of the street and horse-drawn carriages.

"How are your ribs?" Jeanette asked.

Cody laid a paper napkin in his lap. "Much better."

During the past couple weeks the swelling on his jaw had subsided and his black eye had lost most of its discoloration. His ribs and where he'd been struck on the back still hurt when he moved the wrong way.

A young male waiter took their order. Cody decided on a small salad and ice tea. Jeanette selected a bowl of soup and coffee.

She looked at him. "Is there something on your mind, Professor."

"Why?"

"You don't hide things very well."

"I think I've came to a decision."

"Care to tell me?"

"I got to think it through."

"The last time you had an idea, you got beaten up."

He laughed softly.

Jeanette asked, "What about you, Professor?"

"What about me?" He asked, stabbing his fork into the salad.

"How long are you planning on staying?"

"I came down to help Mitch. I'm not leaving until we get things cleared up."

"You think that's possible?"

"I'd like to think so. But Gear isn't going to give up easily. I've learned that."

"It makes me furious to think of them getting the marina."

The waiter brought their order.

"They haven't," he replied, "got it yet."

"Seen Shawna lately," Jeanette asked, trying to sound casual.

"Not since the last time," He referred to her visits during the few days he was recuperating. She had brought soups and a casserole.

"Why do you ask?"

"Oh, no reason."

He knew, of course, why she asked. It has been obvious to Cody that Jeanette had not liked Shawna visiting him.

Their conversation became perfunctory as they watched tourists and horse-drawn carriages pass their window. Cody saw that Jeanette had finished her soup.

"Shall we go?"

"I'm ready, Professor."

When they got back to the marina, they found a crowd had already gathered. Everyone looked hungry as they cast expectant glances toward the grills.

Earlier, the cooks had fired up the welded 50-gallon drums. When the coals were ready, they sprinkled the steaks, chops, and chicken with condiments and brushed them with a special sauce concocted by Mitch.

With the food almost ready, a pall of smoke hung over the fires and aromatic odors infused the afternoon air. The cooks looked like demons in a Faustian hell as they wiped their eyes, turned the meat, and drank beer.

Cody got himself a beer, and by the time he finished it, impatient guests had lined up at the grill, sniffing the odors, and staring at the food with anticipation.

"Reminds me of the military," Rocky mused as he joined Cody.

Tony and Maria Pazzio sat apart from the others. Water Rat staggered happily among the crowd with a can of beer in each hand. Gus Johnston tried to get free of his wife to join the fellows at the beer kegs.

Cody saw Shawna talking to another woman. When she looked his way, she smiled and waved. He thought she looked nice in her dark, well-fitting slacks.

With the increasing consumption of alcohol, conversation grew more animated; laughter floated up on the hot humid air. And now that the food was served, paper plates became balanced precariously on laps, while ants feasted on food spilled on the ground. Jugs waddled about, pausing occasionally to juggle or poke the ribs of an unsuspecting male.

Cody had just opened another beer when Larry MacIntosh and Lori approached.

"Hello," Cody greeted him.

"By gosh," Larry remarked, "this is something. If it weren't for the way everything was done. I mean if you can be sure all the time. It just occurred to me, wondering about health and sex, if going into details is really worth it, if no one can fathom the difficulties inherent in a near death experience."

Tactfully, Cody said he just wasn't sure, but would think about it. MacIntosh smiled and added, "There are strings attached to everything just as the mountains are anchored to the earth." Lori took his arm and led him away.

"Cody, how are you feeling?" Shawna had left the woman she had been talking to and approached Cody.

"Much better."

"You look a lot better."

"Anything new on the complaint?"

Cody had made a formal complaint over the beating.

"Not that I know of." She made a futile gesture. "Nothing will be done about it.'

"I really appreciate your coming by while I was laid up. And especially," he added, "the food."

There was a shorter line at the grill. Cody suggested they get their food now.

Jeannete saw them together. When Cody waved, she threw him a dour look.

They made there way down the line, trying not to breath in the heavy smoke rising from the cooking steaks, chicken, and ribs. With their plates loaded, they found two chairs. Cody mentioned to Shawna he'd received a call from Congressman Gannet a couple days ago.

"You got a call from Gannet?" She asked, surprised.

"He says he wants to talk with me. He claimed he liked the article in the paper."

"I find that interesting. Gannet has yet to declare himself on environmental issues."

"Whatever his reasons, I wish you'd join us. He suggested lunch Monday at the Scarlet O'Hara. I'd feel more comfortable with an attorney." He looked at her. "Not just any attorney." The Scarlet O'Hara was a local restaurant near the college. It catered mainly to students and tourists.

Shawna put down her leg of chicken. "I'm flattered, but why do you feel you need an attorney?"

"I haven't had much experience with politicians. I want you to hear what he has to say."

"Monday is a busy day, but for this I'll get away." She paused. "Why don't I pick you up?"

Their conversation was interrupted by the wail of sirens, as two police cars sped down Riberia Street. The scene wasn't unusual. Police often swept into Lincolnville to conduct drug raids or to respond to domestic disturbances. However, when the two cars careened into the marina, it caught everyone's attention.

"Now what in the world is this all about," Shawna exclaimed. She and Cody watched the cars skid to a stop by the bridge that spanned the narrow inlet to the picnic area.

Cody shrugged. "I guess we'll soon find out."

One of the patrol cars belonged to the local cops, the other, the sheriff. Four officers from each piled out and fell into a lineup.

Several of those at the cookout laughed and made derisive remarks as they recalled their days in the military. One old Geezer yelled, "Hey, snap too, you mothers." His companions chuckled and clapped him on the back.

Two groups of officers, forming a double column, started across the narrow bridge.

Pazzio observed the intrusion with a dark look.

Instantly the two files of officers slammed into one another, which threw their approach into chaos. The bridge wasn't wide enough for both columns.

Sheriff Banks' voice boomed first. "All right, men, get your act together."

At the same time, chief Riker shouted a similar command. The two groups, looking shamefaced, halted, uncertain what to do.

Banks and Rider conferred briefly and apparently reached an agreement, because Riker's police waited while the sheriff's men trotted into the picnic area.

When both groups had crossed the bridge, they fell into line with hands on their guns. Riker and Banks both shouted together, "Parade rest."

Cody almost laughed, but there was nothing funny about what was taking place.

Mitch walked up to Riker and Banks. "What in hell is this all about?"

The two officers faced him. "Got a complaint about a disturbance," Riker replied.

"Come on," Mitch protested, "there's been no disturbance here."

"You calling me a liar?" Riker snarled.

Mitch gestured toward the two patrol cars and the phalanx of officers. "You need a goddamned army to check on an alleged disturbance?"

"Never know what you might run into," Banks replied, glancing malevolently at the crowd.

Mitch shook his head. "I don't believe this. Who complained?"

Neither officer answered.

Sheriff banks turned to his men and ordered, "Give it a walk through." As the officers moved into the crowd, chief Riker issued the same command.

"I don't like this," Mitch protested. Again both officers ignored him. "How come you haven't found who broke into my office, and how about those guys who beat up Cody?"

Sheriff Banks faced Mitch. "You telling us how to do our jobs?"

"Maybe someone better," Water Rat shouted.

The sheriff and police chief exchanged glances. "You think you know so much?" Riker growled. One of his men grabbed a handful of Water Rat's ample hair and yanked his head back.

Cody moved quickly as he grabbed the officer's hand. Immediately, several cops surrounded him. The other officers moved in. Jeanette faced the cop holding Cody. "Take you hands off him," she commanded. One of the other cops grabbed her and pushed her roughly aside.

Instantly, Gus Johntson, Rocky, and Jake Hooper moved to her aid. Larry MacIntosh joined in. He grabbed a cop, but was quickly thrown aside. Fists flew.

In the skirmish, several of Riker's troops received cut lips and bruised cheeks. Gus, a better fighter, didn't get hit, while a club struck Rocky, on the side of his face, but not before he had hit one cop in the eye. Jake Hooper had wrestled one of the sheriff's men to the ground.

Pazzio had been watching the fracas with a grim expression, but remained seated. Suddenly he was on his feet.

A couple of the sheriff's men had grabbed Jeanette. One took hold of the front of her blouse.

"We better do a strip search," he said, leering.

Before he could proceed, Pazzio locked a hand onto the officer's wrist. The man looked up into cold eyes and a face with the dark expression of death.

"You interfering with police work?" the cop managed, but before he could say another word, Pazzio released his wrist and clamped his hand over the cops larynx, cutting off his breath. The man' eyes bulged and glazed over. He tried to wrestle Pazzio's hand away. His knees buckled. As he sank to the ground Pazzio released him. The second cop drew his gun.

"I would think twice about that," Cody told him.

The angry boaters surrounded the cops, who now faced a crowd of men armed with whatever they could lay their hands on, two-by-fours, rocks, sticks of wood; a baseball bat. They had consumed enough alcohol to be mean and reckless.

Two shots sounded simultaneously, as Riker and Banks fired their guns into air.

"That's enough," Banks shouted.

"Okay, Okay," Riker shouted. "Break it up." He ordered his men to get back into ranks.

As the officers and sheriff's men retreated, Cody saw Pazzio leading Jeanette away.

"That was a classic case of the cops acting out of control," Shawna said. "Mitch has cause for a complaint."

Water Rat stumbled after another beer, oblivious now to everything but his destination. The others shook their fists and weapons and shouted at Banks and Riker. They were bold in the face of the officers' retreat.

Chief Riker halted at the bridge and turned to Mitch. "Okay, you listen up. I don't want to get any more complaints. So keep it down. Next time I'll close this damn thing down so fast you won't know what happened until the next Fourth of July."

Mitch's face was scarlet with anger. "You'll get a goddamn complaint about this."

"So complain," Riker retorted, and left to join his men.

Cody, as well as everyone else knew the whole thing had been staged as another effort to intimidate Mitch. As Cody watched the officers drive off, he thought that if he'd had any doubts about the action he planned, he didn't now.

With the officers speeding back down Riberia Street, everyone angrily discussed what had happened. Those who hadn't eaten got into line for food; others replenished their drinks.

A sudden deafening clap of thunder broke over the picnic. Instantly, a sheet of rain swept across the Sebastian River. During the melee, no one had noticed the threatening storm.

Amid screams and curses, the picnickers raced for cover. Within seconds the area was empty, except for the cooks who frantically covered the food, while raindrops exploded on the hot metal drums. Water Rat, unaffected by the downpour, reached into a barrel and withdrew another beer.

Cody shouted to Shawna and Jeanette to run for his boat. They were soaked before reaching it. Shawna's hair streamed down into her face, as Cody helped her aboard. Jeanette yelled through the downpour that she was going to her boat to change. She'd be right over.

Aboard the Columbia, Cody handed Shawna a bathrobe and told her to use the head to change. While she did, he put on dry slacks and shirt. When Shawna came out, he hung her things on a line strung across the cabin.

"While we're drying out we might as well enjoy ourselves," he said.

He mixed drinks while lightning lit up the interior and thunder boomed and crashed. The wind-driven rain beat against the Columbia. Jeanette came over, carrying a towel over her head, and the three sipped their drinks.

"Well, this is nice and cozy," Jeanette said, sitting on one of the bunks.

Shawna smiled. "I guess you can't say Mitch doesn't throw a good Fourth of July shindig."

"Replete with fireworks," Cody replied.

"As I said earlier," Shawna put in, "Mitch has grounds to file a complaint, but I don't believe it would do any good."

Cody recalled the look on Pazzio's face. He couldn't remember seeing such hatred. Pazzio was not, he concluded, a man to fool with.

As he mixed another round the storm bombarded the marina. Both women wondered by his expression what Cody was thinking as he took a long swallow of his drink and thought about his plan.

CHAPTER THIRTEEN

Shawna found a parking space a block from the Scarlet O'Hara.

As they walked beneath spreading oaks, Cody wondered about this meeting with Gannet. Although the man mentioned liking the article in the Record, he gave no further hint as to what he wanted of Cody.

Students and tourists crowded the two-story restaurant as Shawna and Cody entered and looked for Gannet. Cody had no idea what he looked like.

"Over there," Shawna remarked. She pointed toward a table by a window. Cody saw a man in his mid-thirties, wearing a yellow short-sleeve shirt. He had dark hair and a pleasant face. Gannet rose and smiled as they approached. Taking Shawna's hands, he kissed her cheek. "What a pleasure to see you again, Shawna."

Cody wondered, as Gannet turned to him, if the guy was genuinely sincere or a practiced political charmer. Cody took his out-thrust hand.

"Dr. Cody, it's nice of you to come. I've been looking forward to meeting you."

Seated, Gannet asked if they'd like anything to drink. Shawna said she'd take a Coke and Cody said the same.

"We'll have three Cokes," Gannet told the waitress, and as he picked up the menu, mentioned that the hamburgers were very good, adding that actually everything was excellent.

He'd obviously eaten here before.

After the waitress returned with their drinks and took their order, Gannet turned to Shawna. "Tell me," he asked, "how's Mike doing?"

Shawna's boss, Mike Reese was the district attorney.

"He's doing just fine."

"Mike's a great guy," Gannet said, "But I can certainly imagine things get hectic in his office."

He turned to Cody. "I'm also a lawyer. I practiced here in St. Augustine before getting into politics. I grew up in a little town in Georgia."

Cody saw at once that Gannet, like most politicians, liked to sneak up on a topic, sniffing the wind ahead of time to determine the best approach.

"I gather you haven't been here very long," Gannet said to Cody. "Of course it's a tourist town," he said not waiting for Cody's reply. "Quaint in its own way."

Cody did not think it appropriate to mention the undercurrent of racism outwardly disguised as tolerance of blacks.

"I wish," Gannet said, "that I could spend more time here, but my responsibilities in Tallahassee don't allow it." He paused, smiling, and then leaned forward, resting his elbows on the table, his expression suggesting he was ready to get to the point.

He looked directly at Cody. "I liked what you said in that interview in the Record. You didn't pull any punches." He paused. "It helped me make a decision. But first, what else have you written?"

"A few articles for referred journals and I have an article due to be published in a month or so in a national magazine."

"Now about my decision. I have decided to take a stand against the take over of wetlands by developers."

Cody raised his eyebrows.

"So far I haven't declared my position on the environment. This might surprise you, but in politics it isn't always that the right way is best, or should I say, the most expeditious."

Their lunch arrived and Gannet beamed at his hamburger.

After Cody's first bite, he had to agree with Gannet's observation. The meat was thick and juicy, the lettuce and tomatoes fresh. The plate was also loaded with crisp, golden fries.

"I'm curious," Cody ventured, when Gannet seemed lost in his hamburger, "why you want to talk to me."

Gannet wiped his mouth. "I've been thinking of introducing a bill or a series of bills, depending on what it would take, to protect the environment against unplanned and irresponsible development. However, I don't feel I have the necessary expertise with regard to the environment. I'm afraid I have a lot to learn regarding the impact of developments on our ecology." He studied Cody. "I'd be interested in your help."

Surprised, Cody glanced at Shawna. Her slightly arched eyebrow conveyed this might be interesting.

"You're suggesting," Shawna said, "that Cody act as a consultant."

"That's one way of putting it." He hadn't taken his eyes off Cody. "You could help me put the legislation in proper shape from the standpoint of ecological facts." He paused, a slight smile tempting his lips. "The pay wouldn't be all that great. I have a limited budget."

Gannet closed his eyes and massaged his temples. Then he exhaled a deep breath and said, "I might as well admit it, I have sat by timidly, or perhaps I was vacillating, as I watched contractors, developers, and their lawyers browbeat the legislature into caving in to their demands and my legislative brothers selling out to the developers."

Cody was surprised at the man's candor. However, while he saw Gannet's proposal as a chance to fight for the environment and make a few bucks, he had some serious misgivings about what the man had just said.

"It's an intriguing idea," Cody remarked. "I understand your desire to protect the environment. I agree something has to be done. However…" Cody put down his hamburger. "Do you mind if I'm candid?"

Gannet leaned back, his attention focused on Cody. "I wouldn't want you to be anything but."

"Before I explain," Cody began, "I think you should be aware of what's been going on concerning Condo Developers, the firm owned by the Gears."

Gannet listened closely as Cody told him about the potential destruction of the environment where the condos were going up, the break-in of Mitch's office, the beating, and the raid on Mitch's Fourth of July picnic. Cody said he was sure the Gears were behind all of it.

"But we can't prove it," he added.

Shawna spoke up. "The Gears have created a power structure here in St. Augustine that includes the police, city commissioners, and a bunch of good-old-boys. When you have the mayor, the city attorney, the commissioners, not to mention a state senator in your pocket, you got a formidable group."

Gannet shook his head. "I guess I have spent too much time in Tallahassee. I knew, of course, that Vic Blanchard had an interest here. I assume he's the senator you're referring to."

He turned to Cody. "However, I still want to hear what you think about my proposal."

Cody glanced at Shawna, then at Gannet. "I don't think it'll work."

Gannet looked surprised, but said nothing.

Cody continued. "Do you think you can get enough support to push through the type of legislature you're proposing? There are powerful interests that will fight it. Besides, we have enough laws, rules, and regulations on the books for protecting the environment. Do we need more? The Florida Department of Environmental Regulation is supposed to enforce those laws. The corps of engineers is supposed to help protect the environment, and there's the fisheries and wildlife people, and water management districts, and they all are supposed to protect our ecology."

Cody sipped his coke. "I don't think we need more laws and regulations. We need stricter enforcement. Throughout the state wetlands are losing ground to condos, malls, you name it. If the interests are powerful enough, the wetlands are taken. Money and power have a great influence.

"Perhaps you've heard what some of the more extreme environmentalists have been doing. I'm referring to ecotage. It involves people taking the law into their own hands by sabotaging projects considered harmful to the environment."

"Are you suggesting I endorse lawlessness?" Gannet asked.

"Of course not. Although that recourse has been effective, at least to some extent. However," Cody added, "in regard to your offer, I would be willing to review whatever you propose, strictly from an ecological point of view. Still, I don't think you can pull it off. At least not as you've proposed it."

Gannet reflected a few moments. "You may be right. I'm going to have to think about it. You alluded to the radical environmentalists. Care to elaborate?"

"I used to think they did more harm than good, but now I'm not sure. In view of the stance the administration in Washington is taking regarding the environment, perhaps drastic action is necessary."

"Do you condone breaking the law?" Gannet asked.

Cody hesitated. "I am not suggesting anyone break the law. "But there's two sides to this issue. Take, for example, the over-reaction of the timber companies? The son of the owner of the Lancaster Timber Company in California broke the nose of a woman protester. That was stupid and irresponsible. And it's just as irresponsible when the forest service, buckling under timber company pressure, closes sections of our public forests, illegally, I might add. Attempts to use the courts to stop developers from destroying our environment are nothing short of frustrating."

"Gannet spoke up, "I have to use the means available—within the law."

"Of course." Shawna said. "Unfortunately, as Cody pointed out, we need better enforcement of the laws designed to protect our environment."

"I've spent a lot of time," Cody said, "fighting, mostly in academia, for the preservation of marshes, wetlands, and natural habitats. But with little success." He threw a glance at Shawna. "I don't believe our environmental problems can be solved simply by legislating car exhausts and factory emissions, or by halting the cutting of our rain forests. We got to get everyone into the act. We have to make people believe it does more harm than good to pollute and destroy our wetlands. We must make them understand that it is for their own good."

"That's a tall order," Gannet said.

"Unfortunately," Cody continued, "our society thinks only of satisfying immediate appetites and to hell with the consequences. Let someone else worry about the environment. Which of course is not only shortsighted, but also self-destructive. The way I look at it, money, politics, and greed are the three horseman of wetland destruction. We're in a fast-food culture with a channel surfing mentality. Do you think the millions of people who view idiot talk shows and the Jigaboo sitcoms give a damn about the environment?

"What we need," Cody added, looking directly at Gannet, "is a widespread spiritual awakening. An understanding of why we're here on this earth and what obligation we owe to it. We need to take drastic action, to get the right kind of attention directed at our dwindling wetlands."

He glanced at Shawna, thinking she must really think he had gone off on a tirade.

Both were looking at him.

Cody continued. "As Shawna knows, I've been writing to try to awaken people to our environmental problems with a focus on our diminishing wetlands. I hope to reach a much larger audience than my jaded students or those who read refereed publications."

He paused as he sipped his Coke, then looking directly at Gannet he leaned forward on his elbows. "There may be a way to approach our problem. Bluntly put, why not have the state and/ or local governments buy up wetlands before the developers get their claws into them?"

Gannet looked surprised. "You'll have to be more explicit."

"It's an idea that has been in the back of my mind for a while and it could be where you would play an important role. Bonds could be issued for what I call a land acquisition program. It should be given a name, such as Preservation Florida, or Saving our Environment, or whatever."

Shawna looked at Cody with both surprise and interest. He had not mentioned this before.

"The main idea is that the state appropriate money that would be used to buy up environmentally sensitive lands. Decisions would have to be made with regard to how Florida will manage the land it has purchased, how will it prioritize land purchases, and what kind of recreational uses could be allowed on such lands."

Cody finished his Coke. "I can imagine about four main objectives to the program: one, to put greater emphasis on managing the lands Florida buys; two, place a higher priority on purchasing lands in urban areas and areas surrounding aquifer recharge areas and along our waterways; three, allow the water management districts to invest in cleaning up our waterways; and four, allow Florida to take significant steps towards truly developing a system of greenways and trails for recreational use by our residents. The main thrust of all this is, of course, to protect our precious wetlands and other areas that are environmentally sensitive."

Gannet didn't speak for several moments. Finally, "The idea intrigues me," he remarked, sitting back. "True, there's a lot of work to be done on this, but I think it has possibilities. We'd have to convince the legislature that there's merit to this. Those guys in Tallahassee want something for their money. Nevertheless, I got to give this some thought and see what I can do. It just may be the way to help preserve our vanishing wetlands. Incidentally, I'd like copies of anything you've written. They might help me with this."

"I don't want you to think," Cody said, "that I am criticizing you personally or your concern for the environment. We need more legislators like you. But, I just don't believe more laws on the books is the answer."

"You could be right. And I wish you every success, with your writing. Keep it up. We need all the help we can get in our fight with the developers. I'll think about everything you've said."

Gannet turned his attention to Shawna as a signal he was ready to leave. "It has been a pleasure seeing you again, Shawna." And, extending his hand to Cody, he got up. "I'll be in touch."

They went outside, and when Gannet left, Shawna suggested a walk along the quay. They crossed the Avenida Menendez, where the horses and carriages waited for tourists.

Cody looked out at the boats anchored in the waterway.

Shawna put her arm through his. "You have any plans?"

"Plans?" he asked.

"For this evening."

"Not really."

"How about dinner?"

"That sounds good. Where would you like to go?"

She squeezed his arm. "My place."

CHAPTER FOURTEEN

Cody sat waiting in the marina cafe. It was nine o'clock and getting dark. He had picked this evening because there would be no moon.

Although he felt confident his plan would work, he nevertheless felt a lingering trepidation. Perhaps, he wondered, it was because he was still somewhat surprised at his decision. But, he reasoned, what were the alternatives?

Cody conceded that if his plan were successful, it would only temporarily set back Gear's construction. Therefore, he cautioned himself against believing it would create a sudden ground swell against the project. However, he saw his plan as the only feasible way to deal immediately with the threat to both the wetlands and Mitch's marina.

Earlier, when he told Rocky what he had in mind, Rocky had nodded without showing surprise and insisted that he be part of it. Which, of course, was what Cody hoped. With Rocky involved, it would be easier to recruit others, since Rocky carried considerable weight around the marina. Nothing had been said to anyone else.

"Kind of glad you're doing this now, Rocky had said, "because Millie and I are leaving soon. Heading for Trinidad."

Cody would miss the Brinsons. So would a lot of others.

After his talk with Rocky, Cody spend a couple days selecting those he thought would be willing to help out. Each was informed their participation was strictly on volunteer bases. No hard feelings if anyone declined.

The screen door to the cafe swung open and Rocky entered, followed by Gus and Marion. Behind them came Jules Rodrigues and Jake Hooper.

Cody recalled that Jules had said he'd also be leaving soon. Like the Brinsons, Jules was among those who had completed refurbishing their boats and was eager to get back on the high seas.

"Hear you're leaving soon," Gus said to Rocky. And added, "Me and the misses are heading out too. The other direction. Hawaii."

As he waited for the small talk to end, Cody felt a twinge of melancholy, thinking of those who were leaving and wondering if he would ever see them again.

"First off," Cody said, "I want to thank you for coming. As I explained before, there is obviously danger, so if anyone wants out, it's okay."

No one spoke.

Cody recalled that Larry MacIntosh had gotten wind and had volunteered. It had taken considerable effort to dissuade him. Unfortunately, there was no way to keep such a plan secret, and it wasn't long before word had spread throughout the marina. No one, however, mentioned knowing about it.

As Cody spoke, he saw the expressions of determination in the group seated with him. He knew they hated what the Gears were doing. The environment was precious to them. They lived in it. They respected it, and had developed a symbiotic relationship with it.

"It'll be darker in another ten minutes," Cody said. "We'll start out then. I have the corrosives and the spikes and hammers. If all goes well, we'll put a few chain saws and 'dozers out of business."

Cody did not want Mitch involved, but he had to tell him what he planned, because Mitch would provide what was needed: a chemical to jam up the gas lines of the bulldozers, a couple chair saws, and hammer and spikes.

"I hear," Marion put in, a smile playing about her mouth, "that tree spiking raises hell with chain saws."

"I can well imagine," Jules remarked.

"And it'll take time to dismantle the diesels to clear out the corrosives," Jake added.

"Let's hope," Cody said, "that our efforts will make a statement. At least the Gears will know they can't expect smooth sailing with their project.

"We'll break into two groups," Cody continued. "Marion, Rocky, and Gus in one, and Jules, Jake, and me in the other. The first immobilizes the 'dozers; the second will do the spiking."

"What about noise?" Marion asked.

"There will be some, but we hope it doesn't draw attention," Cody replied.

After several questions, they sat waiting for darkness.

When Cody went to the door and looked out, he said it was time to get going. To Rocky he said, "I never thought I'd resort to this."

"There are a lot of unexpected twists in life," Rocky said. "I hadn't expected to be living on a 32-foot boat and sailing all over the world. But here I am, and with a wonderful woman. I finally found what I was looking for."

Cody reflected a moment. "I guess a lot of the people here have found what they're looking for. Unfortunately, a lot of people haven't. They move through life forever searching."

As Cody anticipated, no workers were at the site. Although he hadn't seen Ben lately, he figured he was probably out shrimping. Ben was the only man at the marina Cody didn't trust.

Regardless of what they accomplished tonight, Cody again cautioned himself against expecting much of a public response. Shawna, who knew nothing of what Cody planned, had mentioned meetings of citizens angry over the treatment of the environment. Nothing had come of these. Judge Klein's daughter, Sandra, had stirred up some young people against the Gears' project, and had got a small group to march with placards in front of the San Marco office.

Cody glanced at Gus and imagined the ex-marine was looking forward to the night's adventure. Marion Talbert was perhaps recalling her part in the rebellious sixties.

When they had entered the site, Cody held up his hand for everyone to halt. He wanted to be sure the way was clear. He then motioned them to follow.

Cody looked at the huge bulldozers and reflected they looked like incarnate monsters in nocturnal hibernation. He pointed toward the trees tied with orange ribbons. The ribbons indicated these trees were to be cut down by the clearing crews.

"Take those first," he said to Jules, Marion, and Jake.

"When we're finished, we'll meet back here and return to the marina. If anyone sees or hears anything suspicious, quietly warn the rest of us, if possible; otherwise, get the hell out of here."

As they spread out, Cody smelled the dank odor of decaying vitreous being devoured by microorganisms and the sea-salt smell drifting in from the waterway. He imagined the life of the wetland, silent, but now caught in a death trap of destruction, and felt a surge of anger over the desecration awaiting this wetland.

He pushed his anger aside as he turned his attention to a marked tree. Before driving a spike, he glanced back. Gus had climbed onto one of the diesels and was unscrewing the fuel cap. Rocky handed him the contaminant. Marion Talbert was climbing onto a second 'dozer.

The sound of a hammer striking a spike signaled Jake Hooper had started on his first tree. He turned to Cody and grinned as he gave thumbs up. Jules Rodrigues started on another tree.

By driving the spikes deep into the trees, they wouldn't show. When the power saws cut into the trunks, their chains would shatter. Cody hoped the workers would not be injured, but he couldn't worry about that now.

Apparently, Cody reflected, no one had heard the hammering or saw the figures scrambling up the diesels. He thought about the Gears' reaction when they heard about this night's work. He permitted himself a brief chuckle.

It was brief.

The explosion of light blinded him. In that moment of surprise and confusion, he saw the silhouetted forms caught atop a bulldozer and Jules and Jake about to attack a tree.

He tried to see who was confronting them, but could not, because of the lights.

"You all just stand where you are and don't move."

He recognized chief Riker's voice. Damn, They'd been betrayed.

"Drop all that goddamned stuff," Riker commanded. He then shouted at those on the 'dozers to climb down. "Right now!"

Cody imagined the man's bloated face, dark mustache, and pot belly. He suspected the idiot had drawn his revolver. And as Riker came out of the darkness and halted with legs apart, the light behind him, Cody say he had indeed drawn it.

"You can put that gun away," Cody told him.

"You can shut your mouth, boy," Riker replied. Under different circumstances Cody would have seen humor in their actions as the officers came forward, visible now behind their hand-held lights, They moved like hunters stalking dangerous prey, a phalanx of a swat-team, guns in hand, stealthily closing in on vicious criminals,

Riker shouted again. "Keep an eye on these bastards. Don't let any escape."

Christ, Cody thought, how in hell could anyone be that stupid?

As Riker approached, he told Cody and the others to put their hands over their heads. He then ordered a couple officers to collect the hammers and spikes and fuel contaminants. Riker was in his glory, a smirk of self-satisfaction on his face.

"You shit-head radicals should be strung up by your balls," he cried. Then looking at Marion Talbert with a grin, added, "which in your case might present a problem." Several cops chuckled.

"That kind of talk isn't necessary," Cody warned him.

The blow from Riker's fist knocked Cody backward.

Instantly, Gus, chopped the chief's wrist. Several cops rushed forward. Riker ordered them back.

"Fuckin' karate expert, eh?" Riker shouted as he raised his revolver and pressed it against Gus's forehead. "You want trouble, skinhead, I'll give you trouble."

"Hold it," Cody shouted, jumping forward to grab Riker's arm. "There's no need for that."

"Of course there ain't," Riker snarled. He struck Cody with the barrel of his gun. Blood ran down his cheek. Cody pulled out his handkerchief, Quickly, Marion grabbed it and held it against the wound.

Gus Johnson glared at Riker. "You're a brave man behind a gun."

Rocky put a restraining hand on Gus's shoulder. "Cool it, Gus." Addressing Riker, he said, "You can get in a hell of a lot of trouble for that."

"No shit," the chief replied, turning his light beam into Rocky's eyes. "You some kind of latrine lawyer?"

"We call it police brutality," Rocky replied.

Riker turned to face his men. "Any of you see police brutality?"

They hadn't.

"Well," Riker resumed, turning back to Rocky, "you listen up, fuckass. You speak when I tell you to. And that goes for the rest of you."

Slowly, glaring at the group, he holstered his .38.

"Now," Riker said, the flashlights catching his smile, "We're all going to go for a little ride." He turned to his men. "Take 'em in."

Four patrol cars swung into view, their headlights probing the night.

CHAPTER FIFTEEN

The city jail was intolerably hot. It reeked of sweat and urine.

When Gus protested the odor, the skinny jailer replied, "Tough. And the air-conditioning ain't workin' either."

The jailor had shoved Gus, Jules, and Jake into one cell, obviously enjoying his authority. He put Cody and Rocky in an adjoining one. Riker personally escorted Marion to a cell in another section of the jailhouse.

They were no sooner out of sight then Marion shouted a stream of obscenities accompanied by a yell of pain. When the chief returned, bloody scratches ran down the side of his face.

"What in hell did you do?" Cody demanded.

"Shut up," Riker retorted, and to the jailer he barked, "Any trouble from these guys, hose 'em down, hear?" He slammed his office door behind him.

"Jesus," Gus remarked, "I'd like to get my hands on that sonofabitch."

Before being thrown into their cells, Cody had demanded to make a phone call, but the chief replied, "You can make it in the morning."

"We have a right," Cody persisted, but Riker ignored him.

Seated on his bunk, Rocky retorted, "That was wrong, denying us a call."

"Unfortunately, there's nothing we can do about it," Cody said. By now, he thought, everyone at the marina must know about their arrest. The wives would be frantic.

He wondered if he should call Shawna. But no, he didn't want to put her in a difficult position. She represented the district attorney's office.

It wasn't her responsibility to spring people from jail. However, she might contact or recommend an attorney to get them out.

Their arraignment would be before a judge, and Cody wondered which one it would be, Henderson or Klein. Shawna told him that Crutch Henderson was an establishment judge and would not be sympathetic. He'd lost a potential profit from a valuable piece of land when a group of ecologists successfully fought his efforts to turn it into a housing development.

Judge Klein, however, would be far more tolerant. According to Shawna he idolized his daughter, who supported the environment and had formed groups of young people opposed to destroying the wetlands. He had raised her after his wife died in a car accident.

"This place stinks," Jules remarked as he took stock of their Spartan accommodations.

"Figuratively or literally?" Cody remarked sardonically. "Gus has already commented on the odor."

"How long you figure they'll keep us?" Jules asked.

Cody ran a hand through his hair. "We ought to be out tomorrow. If we can get a lawyer."

He didn't express his concern how the legal system would treat them, nor did he say anything about where the bail money would come from.

"Shit," Gus exploded, pacing the adjoining cell.

Rocky observed, "Someone turned us in, Cody."

Jake Hooper grasped the bars of his adjoining cell. "We can guess who."

Cody didn't have to guess.

"I never liked the two-faced bastard," Jules remarked.

"You figure Marion is okay?" Gus asked.

"I don't think Riker got very far, "Cody replied. "You saw his face."

Jake mopped his forehead with a handkerchief. "It's going to be a long hot night without air in this stinking trap."

The others mumbled agreement.

Gus looked up at the naked light bulb encased in a protective steel cage. "That stupid light's going to keep me awake."

When they finally made themselves as comfortable as possible on the hard, lumpy mattresses, Cody reflected on the evening's events. While it was true the others agreed to participate, he was responsible. Although he didn't have anyone to answer to, Gus, Rocky, and Jake had wives to worry about; and there was Marion.

Cody pressed the heels of his hands against his eyes. There was no use torturing himself. He couldn't do anything now. Tomorrow he'd make a phone call.

The others had turned their backs to the light and were doing their best to get comfortable.

In the morning, the skinny jailer brought breakfast. It was inedible, probably bought from a cheap cafe that gave Riker kickbacks. The fried eggs were cold, the toast burned, and the bacon rancid.

Gus threw his plate on the floor. "Christ, this would gag a maggot."

At nine o'clock the door to Riker's office was thrown open and a different jailer came into the cell complex. He was fat. His pig eyes surveyed the prisoners with disdain, He stopped at Cody's cell.

"Someone to see you." He stared hard at Cody.

"Who is it?" Cody asked.

The fat man grunted, swung open the door, and motioned with a pudgy hand for Cody to go ahead of him. They went through Riker's office to an adjoining room.

"Hold it," he ordered. The jailer stepped around Cody to open the door, but kept his eyes on his prisoner.

"You afraid I might escape?" Cody asked sarcastically.

"Fuck you," the jailer replied, his jowls quivering.

Cody shrugged.

"Okay, wise guy, inside," He shoved him into the room.

It was small with bare walls. A wooden table and a couple chairs occupied the center of the room. In addition to a bare bulb, the only other light came from a single window.

A woman, with her back to him, stood looking out.

"Here he is," the jailer said and closed the door.

The woman turned. Her blond hair sparkled with flecks of light coming through the bars of the grimy windowpanes.

∗ ∗ ∗

In her small attractive bungalow on Anastasia Boulevard Shawna Gregory listened to the radio as she finished breakfast. Normally, she paid little attention to the morning news, but the first item wrenched her attention from her poached eggs and bacon.

"Last night local police arrested five people caught sabotaging a condominium development off Riberia Street. According to an official source, who wished to remain anonymous, one of those apprehended was Dr. Cody Matheson. The others included …."

She did not wait for the rest, but immediately put her plate in the sink, and hurried into her bedroom, leaving the announcer talking about a tropical storm approaching the Windward Islands.

When she finished dressing, she hurriedly checked her make-up, rushed out to her car, backed out of the garage and sped downtown.

In her office on the sixth floor of the Union Bank Building, she switched on the small radio on the credenza. There was nothing further about Cody's arrest or what had happened at the construction site. The all-news station was covering the latest troubles in Russia.

Her mind raced with possibilities as she looked out the large window at the view of downtown St. Augustine. Rush-hour traffic poured down King Street and along the Avenida Menendez. Another stream went over the Bridge of Lions.

What, she asked herself frantically, had Cody gotten into? Was he all right? She desperately wanted details. What were the charges against him? When would he be arraigned?

The phone interrupted her thoughts. Brad Longstreet, in the adjoining office, wanted to know if she'd finished the research on the briefs she had prepared.

She had; she'd bring them right in.

Brad Longstreet glanced up and smiled as he accepted the manila folder.

Shawna, however, was still thinking about Cody and the newscast. "Have you heard the local news, Brad?" she asked.

He didn't look up from the folder. "Ummmm, no. What about it?"

She tried to sound casual. "There was trouble at the Gears' construction site."

Shawna knew that Brad had connections at the courthouse.

He continued looking at Shawna's folder. "What kind of trouble?"

"I don't know, but the news mentioned Dr. Cody Matheson."

He looked up. "Matheson? Wasn't there a piece in the Record about him recently?"

Shawna nodded. "Since he's, well, a friend, I'm concerned."

"Oh?" Brad replied, now more interested. His look implied lascivious thoughts.

"Any charges?"

"Not that I know of, but there will be, of course." She paused. "I'd like to get him out."

Brad thought a moment, then said, "He makes a phone call, gets a lawyer and, bingo, he's out."

"I hope you're right, but I wish you would look into this. Find out what you can. Would you do that for me?"

"For you? Of course. I'll let you know. We can settle on payment later." His smile blatantly suggested what payment he would like. Brad worked hard to get it on with Shawna.

She ignored the innuendo. "Thanks, Brad, I appreciate it."

"No problem. Anything for an office mate." Glancing down at the folder, he added, "This is fine."

Shawna threw him a brief smile. Brad waved, his eyes remaining glued to the enticing contours of her retreating backside.

In her office, Shawna stood by her window. Since the night she invited him for supper, he had been constantly on her mind.

Then back at her desk, she glanced at the Record. Possibly the story made the paper before it went to press. She found nothing. She glanced at the front page, which contained a feature about hurricanes. Although not interested, but needing to focus on something, she read it.

During the hurricane season, which runs from June through November, weather disturbances move off the west coast of Africa.

As they move into the open waters of the eastern Atlantic they pick up moisture, which is the life-blood of hurricanes.

In the initial stages of hurricane development, an incipient area of storminess occurs. The first sign that nature may be up to her tricks is a cyclonic or counterclockwise rotation of the easterlies.

When given a transfusion of water vapor, the area of storminess may intensify into a tropical storm with wind speeds of at least 45 miles an hour. And if meteorological conditions continue favorable, a tropical storm may develop into a hurricane.

Her mind went back to Cody. Maybe she should call an attorney. She knew several who could probably get him out.

She read on. *The birth of the hurricane occurs when the winds reach 75 miles an hour. This process, from a subtle change in wind flow to a full-fledged hurricane, takes anywhere from a couple days to a week.*

Whether the mating of weather elements produces a full-blown hurricane is always a big question.

Shawna hastily read the last paragraph as the writer closed with the comment that *St. Augustine had been lucky to be spared the brunt of most dangerous hurricanes. This had created an attitude of complacency.*

Despite the admonitions of the Hurricane Center in Coral Gables that St. Augustine is a disaster waiting to happen, the city's citizens trust more in the comparative safety of selective memory than the facts of nature. However, meteorologists agree that the First Coast area is a sitting duck whose luck might soon run out.

She folded the paper and tossed it onto the credenza. She recalled that Cody had laughingly told her that because of the intricate interrelationship of the elements of the weather, the flapping of a butterfly's wings in Hong Kong could create a rainstorm in New York.

"And," he'd added, laughing. "if that's the case, God help us when a condor flies overhead."

CHAPTER SIXTEEN

Cody stared at Erica Gear. Hardly the person he expected to visit him in jail.

She turned and stared at his cheek. "What happened?"

He instinctively brought his hand to the exposed cut.

"Riker".

"He struck you?"

"He held a gun to Gus's head."

"He hit you with his gun?"

"That's right."

"He shouldn't have done that."

"That's an understatement," Cody replied icily.

She motioned to the chair. "Let's be more comfortable."

"Why are you here?" Cody asked as they sat down.

"Ed wants you out of the way." Erica said, dismissing the subject of Riker's action.

"You're here to tell me that? And just what do you mean by 'out of the way'?"

"No, to the first question and, well, to the second, he hates your interfering."

"Too bad."

"I'm sorry about what Riker did."

Cody felt exhausted from lack of sleep and incensed over the treatment he and the others had received.

"Ed doesn't always have the final say," Erica continued.

"What's that supposed to mean?"

"Just that."

"What about the beating? Who had the final say?"

She stared at him a moment. "That was stupid."

"Another understatement."

Her eyes grew soft. "I'm sorry about that, too, Cody. Really. Can we forget it. At least for now? I have something I want talk to you about."

"Look, I got a lot on my mind, not the least of which is getting us out of here. I need to call a lawyer."

"And last night Riker tried something with Marion. He got his face scratched for his efforts."

Erica studied him. "Are you sure? Did you see what happened?"

"Of course not. He had taken her to the women's section."

"But no one knows what he did, right?"

"Marion does."

She sat without speaking a moment, then said, "Right now we have to decide what we're going to do with you and the others."

"We?" Cody asked. "You and Ed? Since when do you two have the authority to decide what to do with us?"

He'd no sooner spoken than he saw the change of expression in her eyes. It was as though he were facing a cold and deadly reptile.

"That's not important, Cody. You are." As suddenly as her eyes had hardened, they softened. "I have a proposition."

"A proposition?"

Cody looked at her sharply, wondering what she was leading up to and what her proposition was.

"I can arrange for you all to be released." She paused. "And I am prepared to offer you a job."

He stared at her. "Better pass that by again."

"Cody," she said softly, "You and the others can go free. I mean it. But you have to agree to what I'm offering."

"I'm listening, but that…"

"Judge Henderson might not think releasing you all a good idea," she interrupted, "when we present our charges against you. What would prevent all of you from sailing out of here? Now listen to what I have to say."

When she finished, he shook his head. She's out of her mind, he thought.

"You got to be kidding," he said finally.

"On the contrary."

"Why? I don't get it." He continued to stare at her. "You get me beaten up and now this."

"Cody, please understand. I did not condone the beating. Like I said that was stupid."

She leveled her beautiful eyes at him while a shadow of a smile crossed her lips. He was still struggling with the incredulousness of her offer.

She had asked him to be the director of the Flagler lab.

"Do you think I'd even consider it?"

"Yes."

"You've lost it, Mrs. Gear."

She pushed her chair back and crossed her legs, exposing a thigh. "I think it's a good idea to have you around. And, please call me Erica."

"This is bizarre," Cody said.

"You need the money. We'll pay you well. You could do research. You will have free rein to study the wetlands and their ecology."

"Have you any idea what you're saying? Christ, I'm against all that you're doing. I hope to hell to stop you from ruining the environment. And now you ask me to go to work for you? You really don't understand where I'm coming from, neither you nor Ed. The answer is no."

Her eyes grew hard again. The return of the serpent. "Let me be more specific. If you take this offer, you and the others can go." No charges will be filed."

"And if I don't?"

She spoke evenly, her eyes riveted to his. "You all stay in here and face charges."

"Goddamnit," he exclaimed.

"I doubt if you'd get out on bond," she continued. "Like I said, not if Judge Henderson presides. He's no friend of environmentalists. Besides, do you and the others have the money for bail? You're risks, Cody. And like I also said, what would stop any of you from sailing away?"

Cody thought of Rocky, planning to leave within a few days, and Gus, who was ready to go. The others would soon be leaving, too. Erica had put him in the position of deciding their fate.

For the moment, he said nothing. Then, "Ed knows about this?"

"Of course."

"And he okayed it?" Cody asked, thinking that was inconceivable.

"He had another idea, but," she said with a slight grin, "I convinced him there was a better way."

"Damn!" Cody exploded.

"It won't be so bad. Think of the possibilities."

"The possibility of everyone at the marina finding out I have gone to work for the Gears."

Cody saw something sinister in her smile. It was like an epitaph to a fate he didn't care to consider. What the hell kind of research could he do if handcuffed to their company and under the Gears' control? He asked her that.

"Let me explain, Cody," Erica said. "You'll be head of the lab. You would advise us on environmental conditions at sites where we proposed to build. Think, for a moment, I doubt if any other firm would have such a set up. It would show that Condo Developers was concerned about the environment. We'd advertise this in new brochures. You would be singled out as the Director of Environmental Investigations. Not only would we get excellent publicity, but you could benefit, too. We have plans for other developments, Cody. Starting with the next one you would have a strong say in how we proceed and whether the designated area would be environmentally sensitive. For instance, suppose we wanted to develop a particular area, you would investigate its ecology and give us an environmental assessment. If there would be severe problems to wildlife, or whatever, we would reconsider. That isn't all. I have another idea which I think you'd really like, but I don't want to discuss it now."

This is absurd, Cody reasoned. Since when had the Gears had a change in attitude toward the environment?

"Suppose you have a cite and I tell you it is not a good place to build because of the environment. What then?"

"We'd listen to what you had to say. That's the agreement. We'd see if there was a way to build around the environment."

Cody didn't believe this, but there was the chance he could get everyone out of jail.

"If you agree, Cody," Erica said gently, "You'll all be out immediately."

He went to the window and gazed out at the brilliant August day. The jail was hot enough, but it was a scorcher outside. It then occurred to him. Suppose he took the offer. Was it possible he might find something that would link the Gears to the incidents at the marina and to their plans for taking it over? Would he be able find out about other projects the Gears had in mind and block them, if they would harm the environment? He might be able to tip Gannet off, if the man could convince the state legislature about buying up endangered land. He didn't believe for a minute what Erica told him.

Finally, with measured words, he said, "If, and I emphasize the 'if,' I accept, do you guarantee the immediate release of the others and that I have free rein in my research?"

Erica smiled.

CHAPTER SEVENTEEN

Late Sunday afternoon Shawna picked Cody up at the marina and drove out to her place. The unexpected release of everyone from jail left her with too many unanswered questions.

In the kitchen, Cody mixed drinks while Shawna prepared supper.

"I can't understand why the Gears didn't press charges," she said. "And I want to know how you got out of jail."

Cody leaned back in his chair and sipped his bourbon and water. "Simple. They think they have me under their thumb."

She turned from the sink where she was paring potatoes. "Care to elaborate?"

He told her about Erica's proposition and added, "I haven't said anything to anyone. I don't want them to know the position I was put in to get them out."

"I have to admit I'm surprised about that offer."

"You can imagine my reaction when Erica laid it out. However, I think you can see the possibilities. And," he added, "it isn't forever."

"There has to be more to this. I have a bad feeling about it."

"I'm not exactly overjoyed, but I intend to check the lab's files to see what I can find."

She got steaks out of the refrigerator and put the potatoes on to boil. Then retrieved a package of frozen peas and after putting them in a pan on the stove, sat down with Cody. "You know they can't hold you to this, so called, agreement. You can tell them to go to hell, legally."

"I thought about it. But, the Gears got Henderson to go along with the release on the condition that I take over the lab."

"But, Cody, it isn't legal. He can't force you to play along."

"I'm sure you're right."

"I'd like to know just what happened at the project site."

Cody told her about putting contaminants in the diesels' tanks and pounding spikes into the trees.

"Had the plan been in the back of your mind when we talked with Gannet?" she asked.

"Sort of."

He looked across at her. "Not changing the subject," he said, "I'm surprised you aren't married? You're a very attractive woman."

She smiled. "Thanks for the compliment. I was, but it lasted less than a year. The guy was a self-centered bastard. I should have seen it, but I didn't. Well," she shrugged, "it's over. You don't know how worried I was when I heard on the radio about your arrest."

"Someone ratted on us. We know that."

"Do you know who?"

"We think so. A guy named Ben, a shrimper. The point is, by disabling the 'dozers and spiking the trees we would have put Gear out of business for a while. I had also hoped it might get people around here motivated to do something about saving the environment. It didn't turn out that way."

"If you find anything in the lab files, make copies. I'll definitely want to see them. However, we have to be very careful not to compromise any data, or ourselves, either."

Shawna put the steak on and then opened the cupboard. "I have a bottle of Sauvignon I've been saving for a special occasion."

He found the opener and after drawing the cork out of the bottle, poured their wine.

Over the meal, they discussed his new position as lab director and the possibility he might find something incriminating in the lab's files.

"It's a long shot," Cody admitted, "but certainly worth a try."

When they finished eating, he helped her clear the table. Shawna put the dishes and silverware in the dishwasher and suggested they have an after-dinner drink in the living room.

She got a bottle of Irish liquor from a cabinet.

"Care for some music?" she asked. She put on a CD and joined him on the couch.

"Great meal," Cody remarked.

"I'm glad you liked it. Incidentally, any nibbles on your articles?"

"Not yet, but it takes time for editors to review all the material they get."

"Of course I wish you luck. However, you know your subject. I can't imagine a magazine not accepting your material."

For a moment they listened to the CD, then Shawna asked, "When will you start at the lab?"

"No date was set, but I figure in a day or so." He smiled, "don't want them to think I reneged."

He thought a moment. "If we take the Gears to court, how difficult would it be to prove their project would be detrimental to the wetlands?"

"I think we could prove it, but it might not be easy." Then she said, "Come on. I enjoyed our dancing at the Palms: let's do it."

With the rug rolled back, they danced.

Finally, as darkness gathered outside, she looked up. "Do you have to go back?"

"I don't have to. No."

She laid her head against his shoulder.

"Good, Why not stay here?"

When the music ended, they finished the liquor, and as though by tacit agreement, they went into the bedroom.

She was undressed and waiting when he came out of the bathroom. As he got in beside her, she turned to him. "I didn't plan this."

"Are you sorry?"

"Of course not, my darling.'

"You're a very attractive woman."

"That's nice to hear. And you're an attractive man."

Her breasts felt warm and soft to his hands. And as he caressed her, she reached beneath the covers for him.

"Oh," she breathed, "you're ready."

"I think we both are," he said, as he reached between her thighs.

"I think I'm falling in love with you, Cody."

He eased himself gently on top of her. "I think the feeling is mutual, sweetheart." He used his hand to position himself into her. She moaned softly while arching her hips to enclose all of him.

"Oh God," she groaned. "That's wonderful."

They made love without hurrying, her hips rising to meet his thrusts.

"God, it feels good," she moaned.

"I can't keep this up much longer," he breathed.

"Don't."

He forgot about the wetlands, the Gears, and the threat to Mitch's marina. He became lost in her warmth and softness, and in the ecstatic release that poured out of him while he felt her gripping throbs as she accompanied him in her climax.

When it was over, he said, "I don't think I have the strength to get up."

She laughed. "Why do you have to?"

"I don't want you to have to explain why a man is leaving your house in the early morning."

She caressed his face with her fingertips, avoiding the still-sensitive area where Rider had struck him.

"I'm not," she added, "some adolescent girl who has to account for her actions. I'll drive you over to the marina on my way to work. So let's not worry about appearances."

When they awoke, the sun was streaming through the bedroom windows.

* * *

"Dr. Chapman?"

It was Monday morning, several days later.

Cody entered the lab to meet the out-going director, having ridden his new Harley motorcycle, which he'd bought from a student strapped for cash. It was a 1975 FX Super Glide Harley Davidson with a 1200cc engine, chrome plated and bright red. Although none of the other professors or administrators had said anything, Cody sensed they did not look kindly on a professor at Flagler wheeling about on a noisy 'hog'.

He found Chapman in the office at one end of the lab. The director had a plump, pleasant face. From behind wire-frame glasses, his small eyes took a moment to focus. He ran a hand over the top of his balding head.

"Yes?"

"I'm Dr. Cody."

As Chapman spoke, his head bobbed, as though re-affirming what he was saying. His mouth worked, suggesting he had something else to say when he finished speaking.

"I've been expecting you," Chapman said, taking Cody's hand in a limp shake, and glancing briefly back at the open drawer of the filing cabinet he'd been rummaging through. "I was just going over the files. Lots of stuff needs to be tossed out, I'm afraid." His eyes opened and closed.

"Leave it," Cody said. He didn't want Chapman to throw out anything.

Chapman seemed lost in thought. He looked around the office, viewing it for the last time. "Well, yes, that's a good idea, I guess. You'll want to review everything." He continued to nod his head after finishing his statement.

Cody got the feeling the man was addressing no one in particular.

"I haven't organized the contents very well. My wife tells me I'm messy. She wonders if all PhDs are that way."

Cody smiled.

"Your training is in ecology." Chapman remarked, "from what Mrs. Gear told me."

"I've been fighting a losing battle to save our wetlands," Cody explained. "I'm a biologist."

"Tell me," Cody said, "just what is the purpose of this lab? I haven't been what you might call well briefed."

Chapman led him out of the lab's office. As he did, he glanced around as though looking for an answer somewhere among the shelves and the bookcases. "Basically, we have a contract with several state agencies to put together environmental impact statements. Saves the state money."

His tone suggested he was going to continue. But he looked up at Cody with an inscrutable expression. Finally, he nodded to unspoken thoughts, while his mouth worked on words that were never spoken.

Cody had strong feeling the man wanted to tell him something.

CHAPTER EIGHTEEN

A couple days later, Cody hopped on his Harley and sped off to the Flagler laboratory. He still had a lot of papers to go through. Jeanette, who witnessed his departure, wondered where he was going. Since he had said nothing to her about his activities, she was curious about what the Professor was involved with.

So far, except for Shawna, Cody hadn't told anyone that he was director of the Flagler environmental laboratory. He did say he was doing some teaching. That would explain his absences. He could not, he realized, keep up the subterfuge forever.

Nevertheless, he felt guilty he hadn't told Rocky and Millie about his new position. Among the people in the marina, he was closest to them, and felt he owed them an explanation. However, he thought it best to leave the impression he was only teaching. At least for the time being.

Occasionally he spent the entire day in the lab. Sometimes just a few hours. He doubted the Gears would question how much time he put in. And as long as he met his biology class, the only class he taught, the college was satisfied.

Not the students, however. Cody quickly got the reputation of accepting only first-class work from his students. And it didn't take him long to learn that the small liberal arts institution set tuition high enough to discourage "undesirables"—never specified—and yet to entice more "desirable students"—also not specified.

He discovered, too, that the school put undue emphasis on student evaluations of the professors. The obvious result was that teachers who demanded hard work were weeded out and the more lenient and less effective remained to perpetuate a permissive atmosphere and poor scholastic work.

Although the students quickly earmarked Cody as a candidate for their blacklist, he wasn't concerned, and continued to demand heavy amounts of homework and graded on student performance.

Because of his light teaching load, Cody was able to spend most of his time in the lab. Chapman had offered to remain on for a few days to help him "get his feet wet."

The man was, Cody quickly discovered, congenial, inoffensive, and absent-minded. He lived in fear of the administration and avoided controversy. Chapman was not going to jeopardize his upcoming retirement by defying either the administrators or the Gears. He looked forward to retiring on his "little farm" in Hastings, a three-acre plot on which he grew potatoes—everyone in Hastings grew potatoes. Chapman's kids were grown and gone.

As soon as Chapman had turned the keys over to Cody and bid the laboratory a final good-bye, Cody began going through the files. It was obvious Chapman's filing system was at best haphazard.

The titles of folders often bore little indication of what they contained. Nothing was in chronological order and even pieces of related correspondence were found in separate drawers.

As he now got to the last drawer of the filing cabinet, he pulled out the folders and spread them on his desk. He felt discouraged over not finding anything that might implicate the Gears. Then opening one folder, he found an unmarked envelope.

In it was a copy of a report he'd read earlier, one that had rankled him. That report had indicated no ecological problems associated with condo development project in the wetland area across from the marina.

This newfound report, however, while similar to the other, drew the opposite conclusions. Serious harm to the fragile ecology, it stated, would result if construction took place.

Cody was puzzled. Why two reports, completely opposite, on the wetland site? Chapman had signed both. What struck him as odd was that while the two signatures appeared identical, there seemed to be slight differences. Perhaps Chapman's idiosyncrasies extended to his handwriting.

He sat at his desk and studied the two reports, but was interrupted by some one entering the lab. Maybe a student wanted to see him or had some work to do in the lab. A moment later Erica entered the office. She wore a figure-revealing dress.

"Hello, Cody," she said in a lilting voice. "How is our new director?"

He closed the folders with the envelope.

"Hello Erica."

This was her first visit to the lab since he took over. "How is everything going?" she asked as she took a chair in front of his desk and crossed her legs.

"No problems."

"I'm glad of that, but there shouldn't be any, should there?"

"I guess not."

"And the teaching?"

"No problem there, either."

"Good."

An inscrutable expression crossed her face, the kind women have when something is on their mind.

"I've really come to ask a favor."

Cody wondered what kind of favor she might want.

"I need your advice," she said

"About what?"

"I'm thinking of buying some land on the Gulf Coast and I want your appraisal."

"I'm not a land appraiser."

Her lips suppressed a smile. "I want you to look over its ecology."

"The ecology worries you?"

"Does it surprise you? Isn't that what you were hired for? To help us make decisions that won't harm the environment?"

He didn't believe that.

"It could be developed, but to show you I'm serious about the environment, I want you to look it over."

"When?"

She smiled. "How about next week, say, Tuesday?"

He thought a moment. He had nothing scheduled, except to work in the lab. His class was on Friday.

"I guess I can make it."

"I'll pick you up at the marina, say nine o'clock Tuesday morning, if that's all right with you."

He was curious about where they were going.

"It's not far from Cross City." She got up to leave. "We may have to stay over one night. Do you mind?"

The idea of being gone overnight with Erica Gear might have ramifications he didn't wish to consider.

She got up. "Until Tuesday." She smiled. She left, but not without an exaggerated movement of her hips as she went out of the office.

Cody locked up the lab and, as he sped back to the marina, he thought about Erica's request. He was suspicious of her motives. Was it he didn't trust her? Most likely, that was it, but, of course, she could be telling the truth.

Back aboard the Columbia, he mixed a bourbon and water and went on deck to enjoy the late afternoon and its eternal sounds. He had just sat down when Jeanette approached from her boat.

"Permission to come aboard, sir," she said with mock formality.

"Permission granted," he replied.

As she settled in the cockpit, she asked, "Where were you all day, Professor?"

"How about a drink?" he asked.

"The usual, Vodka and ginger ale."

He returned with her drink.

"You going to tell me, Professor?"

"Tell you what?" he asked evasively. He didn't want to get into the discussion. He took a swallow of his drink.

"What's so mysterious?" she asked. "You got a woman somewhere?"

"No, there's no woman."

'Well?" she persisted, regarding him with questioning look.

"I'm director of a lab at Flagler. And I teach a course in biology."

"Oh?" She regarded him with both surprise and interest.

He saw that she didn't connect the lab to the Gears.

As he finished his drink, he noticed her expression, Her eyes were soft, inviting, and he wondered if she was thinking she'd like to make love again. As he recalled the time they had, he felt a surge of arousal. She was very desirable.

But there was Shawna.

CHAPTER NINETEEN

The modest one-story bungalow was set back from the street. A field of potatoes lay behind it. As Cody swung into the carport, a St. Bernard lumbered out from the shade next to the house, sniffed, and sauntered back. Dr. Ron Chapman glanced up from the power mower he was working on.

His lips worked and his head bobbed, as mild surprise moved like a shadow across his face.

"Hello, Ron," Cody greeted him.

Chapman put down a screw driver and extended his hand.

"Dr. Cody," he said.

"No need for formalities," Cody said, smiling.

Ron Chapman wiped his forehead with a rag, leaving a streak of grease.

His wife, a plump, pleasant-looking woman, opened the screen door and smiled. She'd heard the motorcycle.

"Would you gentlemen like some ice tea?" Her voice sounded raspy.

"My wife, Elaine." Chapman said, introducing Cody as the new director of the laboratory.

"Oh yes," she said, still smiling.

"If it isn't any trouble," Cody told her.

"No trouble at all," she answered. "It's all made. I was just about to bring some to Ron. Hope you don't mind that it's sweetened. Ron likes it that way." She glanced lovingly at her husband.

When she disappeared, Cody explained he'd come to ask about a report he found, or rather about two reports.

"Reports?" Chapman asked, frowning, his mouth working. His hazel eyes drifted out across his potato field.

If he knew about them, Cody thought, he wasn't showing it. Considering the shape the files had been in, it could be he'd forgotten.

"One of them," Cody continued, "had an earlier date and indicated the construction of the condo developments would have an adverse effect on the ecology of the site."

Elaine Chapman reappeared with the ice tea. She handed one glass to Cody and the other to her husband.

"Thanks," Cody said, "Just the thing for a hot summer day."

She smiled and went back inside.

Chapman seemed to be collecting his thoughts. He held his glass as though he didn't know what to do with it. Finally, his mouth working before he spoke, he said, "You're referring to the reports on the condo project."

Cody said those were the reports.

"Well," Chapman said, "I think I do recall preparing them. We did an initial environmental impact study on the proposed condo site."

"Did you sign the reports."

Chapman went into thought again. Before speaking he took a sip of his tea. "We, I mean I, did an initial report." He paused, looking pained.

"What about the second one," Cody pressed. He was careful his tone did not put Chapman on the defensive.

The former lab director did not reply immediately, but seemed to weigh the question. Or had he lost track of it?

"Yes, the second one," he repeated. "Because Mr. Gear explained the situation."

"The situation?"

Chapman sipped his drink. "About the original purchase agreement."

"Original agreement?" Cody began to feel like a straight man in a comic team.

Chapman's head moved in a way Cody couldn't tell if he was indicating "yes" or "no". He glanced down at his mower.

"The city bought the land from a widow after her husband died. Gear came to the lab. He brought the paper work on the city's purchase."

"For you to look over?"

"More or less," Chapman answered hesitantly.

"Meaning?" Cody asked, Chapman could easily drive one mad.

Chapman put down his drink and picked up a set of old points he had taken out of his mower. After examining them a moment, he said, "According to the original sale agreement, there could not be any harm done to the ecology of the site. The old widow wanted it that way. That was part of the agreement the city signed when they bought it."

Surprised, Cody thought that if this was true, maybe he had something here. "Can you tell me more about this?"

"Well, I read it quickly."

"But you'd recognize it if you saw it again?"

Chapman frowned, "I think so."

Cody was impatient to tell Shawna about this.

"You see," Chapman went on, "Mr. Gear explained that since the original agreement applied also to him, the law prohibited him from damaging the environment where he planned the condos."

"But damnit," Cody interrupted. "that's exactly what he's doing?"

"Well, I was concerned about that, too, when he told me and after I'd read that piece in the paper, the one that quoted you."

"You were concerned?"

"I knew you were right about the area. But I didn't know how much you knew about the purchase agreement or …." Chapman finished his ice tea. "…if you knew about our reports."

"How could I?"

Chapman fell into thought, as he gazed at his dismantled mower. "You never know about those things."

"Oh?"

"If anyone could have discovered there were two reports."

The conversation, Cody thought, had come full circle. "Some one did. Me. And that's what I came out here to ask you about."

"Yes, I understand. Fact is," he paused, his mouth working and his eyes searching for something to focus on, "I half expected you."

Cody fought to control his exasperation. He felt a growing excitement over what he was hearing.

Chapman brought his gaze back to Cody. He seemed to be choosing his words carefully. "Gear explained that since the purchase agreement prohibited harming the environment, he had to find a way around that." Chapman paused to look down at his mower again.

"Did he suggest a way?"

"He said that the original report had to be in error."

"The original? The one you did?"

Chapman nodded.

"But it wasn't, was it?"

"No."

Cody began to see what was coming. "He suggested you do another impact statement?"

"Well, not exactly."

"Oh?"

"There was no need to conduct another survey. He told me to write up another, much like the original, only delete items that indicated the environment would be harmed. He explained this would satisfy the old woman's wishes."

"And so," Cody said, "he asked you to do another that would be in line with the bill of sale? That is, indicate there would be no environmental problems, because Gear was going to comply with the original document?"

"That's it." He glanced again at the points. "You see, I'm seventy and I don't have much of a nest egg put away. I couldn't afford to jeopardize my retirement."

"You're saying Gear threatened you?"

Chapman's head gyrated. "Not in so many words, but I got the message." He stared off into the distance. "But there's something else."

Cody wondered what else there could be.

"Gear said he'd sign the second impact statement."

"Wait a minute. I don't follow. Your name is on it."

"He signed for me."

"You mean he forged your signature?"

Chapman's last statement was unbelievable. Perhaps it shouldn't have been. Gear was capable of anything.

"Anyone else know about this?"

Chapman thought a moment, then shook his head.

"Perhaps we'd better leave it that way."

"I suppose that would be prudent," Chapman replied, his head bobbing and his mouth working.

Cody extended his hand. He was amazed at Chapman' revelations.

"Thanks for your help, Ron. I have to get back. And thank your wife for the tea."

"I will," Chapman replied, looking down at his mower.

Cody turned the Harley around, and as he got astride, the St. Bernard sauntered out once again to sniff Cody's leg.

CHAPTER TWENTY

On Sunday morning they packed towels, a blanket, and sun-tan oil in the Harley's saddle bags. They wore their bathing suits under their clothing. Shawna wrapped her arms around Cody as they roared out of the marina and down Riberia Street. They crossed the Bridge of Lions, and took route A1A South.

The sky was clear except for a few wisps of cirrus. The weather report called for no change in the hot, muggy August weather. The latest tropical storm report indicated that Eddie with winds of 50 mph was continuing its northwest movement. It's present position was one 75 miles north of Puerto Rico. Heavy pounded the island's north coast. With its continued movement, it posed a possible threat to the Bahamas. However, forecasters thought the storm might take a more northerly course and miss the islands.

They flew past sprawling condos overlooking the beaches, trailer parks, and cement-block houses. They caught glimpses of the sparkling Intracoastal to their right and the ocean to their left. They sped past forests of tangle mangroves, cabbage palm, wetlands, salt marshes, and scrub oak, and looked up to see vultures wheeling against the sky. Pelicans soared over the beaches like a formations of bombers.

"Hold on," Cody shouted, as he twisted the throttle. The bike leaped forward with a deep roar, accompanied by a violent surge of power.

They hurtled through a tunnel of blurred foliage, a world that collapsed into acceleration and speed. Emersed in the angry roar of the 1200cc engine, they leaned into a curve. As they did, a car came toward them on the wrong side of the road. Cody swerved onto the soft shoul-der. The bike fish-tailed. In that fleeting moment he caught snatches of discarded newspapers, cans, and bottles that littered the roadside. He saw the two of them crashing into the palms and oaks that lined the road. Shawna closed her eyes and crushed her face against him.

Then they were back on the road. Safe. Shawna felt numb. Cody cursed the driver of the car, and his own damn stupidity for driving so recklessly.

"You okay?" he shouted.

"I'll let you know when we stop."

He felt Shawna's face still pressed against his back until he slowed to turn into an opening that led to the beach.

"This looks good," he said, slowing and pulling off the road.

They took the blanket and suntan lotion and went down onto the beach, where they spread the blanket. Cody rubbed suntan oil on Shawna's back and she his.

Because it was Sunday, there were quite a few people on the beach. Indolent waves broke with the out-going tide.

Cody sat with his arms wrapped around his knees and watched sand-pipers scurry along the beach.

A sloop plied the offshore winds and a shrimp boat lay at anchor with its net arms extended. Cody wondered if Ben were aboard. A fat man sat in a low chair half out of the water like a beached whale, and a scantily clad young couple tossed a Frisbee in a symbolic interplay of sex. A few plump matrons with waffled skins sauntered along the water's edge.

"What are you thinking?" Shawna asked, one hand shading her eyes, despite her sunglasses.

Cody planned to tell her about the trip with Erica, and especially about the two reports.

"A dream I had a couple nights ago, I saw the wetlands on both sides of the marina filled and condos, parking lots, shops, and movie houses. Everything seemed so damn real."

"I hope it wasn't a prophetic dreams," she said.

"If it was, the wetlands are doomed."

Shawna thought a moment. "Well, not necessarily. Not if something is done to change it."

"But what if I dreamed was the future?"

"A possible future."

"Possible?" Cody asked.

"Well, just because you might have seen into the future does not mean that what you saw has to be. It may have been one of several futures. Suppose you had decided not to take me biking today. How would the day have turned out? Would we be here? It is very possible the condos will be built. It is also possible they won't. Maybe your dream showed just one possibility."

"I don't like the idea of my future being planned."

"Neither do I," Shawna said.

"Incidentally, I finished the second article and sent it off to another magazine."

"You're not wasting much time."

"I feel I can reach more people through magazines. I beat my head against the wall in academia, but this is different."

He looked up and down the beach. "Let's hit the water."

Shawna gained a head start and plunged into the waves ahead of him. But Cody caught her as she surfaced.

"Don't go out too far. Sharks," he warned.

"Listen, I've lived here longer than you."

Although they didn't present a significant menace, enough shark attacks occurred each season to make swimmers cautious.

"You can't be too careful." He dove under and came up behind her and clasped his arms around her. He could feel her soft breasts. "Don't get any ideas, Professor," she said referring humorously to Jeanette's nickname for him.

When they got out and toweled, they lay back down on the blanket, Cody listened to the low booming surf and the voices along the beach. A small plane, towing a banner advertising a restaurant, flew offshore parallel to the coast.

A half-hour later Shawna remarked, "I think I've had enough. I don't want to get burned."

They put on their clothes over their now dry bathing suits and walked back to the bike.

"I could stand a beer," Cody remarked.

"There's a place just south of here," Shawna said.

A few minutes later Cody pulled into the parking lot of a small cafe facing the ocean. It had outside tables under awnings.

"How about outside," Shawna suggested. "There's a breeze."

They both ordered beer.

"I have something to tell you, Cody said."

She looked at him questioningly.

He told her about the two reports and his conversation with Chapman. "And I think this is also important. Although I haven't seen the bill of sale, Chapman claims the woman who sold the land to the city, her name was Love, stipulated as a part of the sale that nothing was to damage the ecology in any way."

Their beer arrived.

Shawna looked at him. "What you've told me certainly suggests possibilities."

"I figured you'd think that. Chapman claims Gear signed that second report, but copied his, Chapman, signature."

"Without Chapman's permission?" she asked.

"I'm not clear on that. Chapman's not the easiest man to understand. They may have had an agreement of sorts.'

"That could complicate matters."

"However," Cody remarked, "I should think it pretty risky to resort to forgery, if that's what it was. He is also ignoring what was included in the bill of sale. Don't forget that."

"No, I'm not forgetting that."

"And there's something else."

He told her about Erica's request that he accompany her to evaluate some land on Florida's west coast. "Somewhere around Cross City, she told me."

Shawna looked at him for a moment before speaking. "I wonder what she's up to."

"So do I."

"I'm not sure I like the idea of your going with her, but how else can we find out what it's all about. When?"

"Tuesday."

Shawna changed the subject back to Chapman. "Did Chapman deny signing the second report?"

Cody nodded. "Do you think you could get an indictment or a restraining order based on the argument that the second report is not only incorrect, but a forgery?"

Shawna sipped her beer. "As you know, the procedure is to go after the granting authority. In this case the Department of Environmental Regulation. But the forged document provides a new twist to the situation, in addition to the fact that the Love woman's wishes are being ignored. These two events could give us what we need to go after the Gears."

"I hope so," Cody said, then added, "I'm sure Chapman will attest to the fact he didn't sign the second document. But even so, couldn't you bring in a handwriting expert?"

"Of course. Where are the reports now?"

"I left them at the lab. But" he added, "I think they're safe. I locked them in my desk drawer."

He watched an old woman go by on her bicycle. A few tourists strolled toward the beach. The sun, behind them, laid a lengthening shadow across the sidewalk and into the street. The edges of the awning fluttered in the sea breeze, and the sound of the waves and the cries of sea gulls drifted in up from the shore. Far out gannets dived for fish.

"Let's have another round,." Cody suggested and beckoned to the waitress.

"I'd like to see the reports," Shawna said.

"Figured you would."

He fell silent. Finally he said, "We're an integral part of this planet, Shawna. The idea we are separate from the earth is wrong. And there's the damned conventional wisdom that we have a right to rape its environment."

She laid her hand on his.

"There I go again, running my mouth."

"I don't mind."

"Well, we really don't have the right to ruin our environment or the wetlands. In fact, you harm one you harm them all. Everything is inter-woven. A guy named Sheldrake has an intriguing theory. He refers to morphogenetic fields. Actions spread out through this field, like ripples on a pond. For example, it's been found that after rats learn to negotiate a maze, rats in other places, not at all connected to the original ones, learned to get through the maze much quicker. It's as if learning is somehow transmitted to all rats."

When the waitress arrived with their beer, Cody raised his glass. "To our success, or I should say, to yours."

She tapped her glass against his. "If we are successful, it'll be because of you."

They discussed the possibility of a trial, the Gear's interest in the marina, and the lousy track record Florida had in protecting the endangered wetlands.

Finally, their glasses empty, Cody asked, "You hungry?"

"What have you in mind?'

"Someplace quiet."

Cody paid and they climbed back on the Harley.

"All set?" said over his shoulder above the roar of the engine.

The Harley surged forward as they headed back toward St. Augustine.

Neither wanted the day to end.

CHAPTER TWENTY-ONE

Erica drove. She picked up Cody just before noon at the marina and headed west on route 207. On the other side of Palatka she pulled into a diner's parking lot.

"What kind of land is this you're interested in?" Cody asked when they'd settled into a booth and had ordered sandwiches..

"I've only seen it once. It's situated near a small stream. Mostly it's wetlands."

"Do you plan to buy it?"

She seemed evasive. "Well, I might."

"For what?"

She smiled enigmatically. "I'd rather not say; that is, right now."

He was irritated by her evasiveness.

They were back on the road within forty-five minutes.

She swung passed a Cadillac driven by an old man barely visible above the steering wheel. The country road was not built for speeds of seventy.

"I don't want to disturb your concentration, but aren't you going a bit fast for this road?"

She laughed. "Traffic doesn't bother me."

"I wasn't thinking of the traffic."

Cody turned on the radio. Eddie was still on a northwesterly course with sustained winds of 50 miles an hour. It was located 200 miles southeast of the Bahamas.

"Find some music, I don't want to hear about the weather," Erica demanded.

Oak and palmettos flew by. Brahmas grazed in fields, where Egrets, perched on the cattle's backs, pecked at the bugs and flies.

"How's the teaching?" she asked.

He shrugged. "I'm not used to the type of students that go to Flagler. They're too damned spoiled."

"Are you happy at the lab?"

"I have no problem with it."

She glanced at the speedometer. "I'm going to make you a proposition."

"Oh?"

She smiled. "Not right now, but I think you'll like it."

He knew it would be futile to try to get it out of her as she risked their lives on the country road.

"Do you have anyone special, Cody?"

Her question caught him off guard. "Special?"

"A woman."

That was none of her business. His stomach tightened as she accelerated passed a semi.

A half hour later they entered Cross City. Cody glanced at the usual small-town buildings—stores, gas stations, and white one-and two-story homes.

On the other side of the town Erica pulled into a motel with a weathered sign, "Seaside Motel." Smaller letters advertised a restaurant and bar.

Erica turned off the ignition and got out. "Would you bring my bag, Cody?"

He picked up Erica's small suitcase and his own overnight bag, containing a change of clothes and toilet articles, and followed her into the motel office.

Inside, a short matronly woman greeted them from behind a small counter. She glanced at Cody. Erica told her they had reservations, one room for her and one for her associate. She sounded very matter of fact.

"We'll have time," she said, glancing at her watch, and addressing Cody, "to get our business done."

The sun was dropping toward the distant line of trees as they got back on the road. Twenty minutes later Erica turned down a narrow unpaved road, which ended at a wire gate with a "No Trespassing" sign. They got out and after Cody pushed the gate open, they walked into a large field beyond which lay a wetland.

"What do you think?"

He looked around. The clearing was several hundred feet across and bordered on both sides by thick growths of oaks, cypress, and cabbage palm.

"What am I supposed to think?"

He looked for signs of wildlife. A red-winged blackbird clung to a stalk, while several vultures circled in the distance. A blue heron floated in for a landing in the tall grass. To the west, thunderheads were building against the blue sky.

She took his arm and led him toward the edge of the wetland. An osprey swooped down over the open water and came up with a fish.

She stopped, still holding his arm. "Isn't it beautiful?" He had to agree it was, but he still had no idea what she planned to do with it.

"First off, I'll build a road from the gate to the lab."

"Lab?" Cody asked.

"That's part of the surprise."

"You've lost me."

"The property extends a little beyond the woodland on either side." she said pointing. "A few trees would have to be cut and the land leveled over there," she pointed again. "I wanted to see what you thought."

He looked around. "I can't say what I think."

"The laboratory would go over there," she nodded to the right of the clearing.

He stared at her.

"Like I said, that's my little secret."

"I suppose in your own time you'll explain what this is all about," he said.

"Of course."

"When?"

"When the time is right.

Although not satisfied, he decided not to pursue it as they walked back to the car.

When she swung into the motel parking lot, the thunderheads dominated the sky.

She said, "I'm in the mood for a drink and I'm starved. How about you?"

The restaurant was small, clean, and modestly decorated. Paintings of sailboats hung on the walls. Several people sat at the bar. Taped music came from ceiling speakers.

Erica pointed at a table. "Let's take that one."

The waitress in a short dress and sneakers took their drink orders.

Shawna reached across and patted his hand. "I know you're curious about all this, Cody."

When their drinks arrived, Erica raised her Gimlet. "To us,"

He barely acknowledged her toast as he downed half the Old Fashioned. He couldn't help think how ridiculous to be sitting in a motel restaurant with the wife of a developer who he, Cody, was bent on stopping from ruining the environment?

As she put down her drink, an odd smile toyed with her lips. "Order another round, darling."

He beckoned to the waitress.

"I find you quite irresistible, Cody. You know that. And for what it's worth, I'm rather well off financially. I own a good share of Condo Developers. Ed is president, but he doesn't make all the decisions."

"So I've heard before," he remarked sarcastically. "But I thought we were going to talk about land."

"Do you like surprises, Cody?"

"Depends on the surprise."

The waitress brought their drinks.

The music of her laugher carried across the room. Several heads at the bar turned.

"I've given this considerable thought, Cody."

"I'm glad," he said, not hiding his irritation.

She put down her glass. Her eyes were warm, expectant. "The property is for you."

"I beg your pardon?"

She leaned forward. "I'm setting you up with a laboratory. It'll be all your own. You can continue your wetland research." She watched for his reaction.

Surprised, and not knowing how to immediately respond, he continued to stare at her. Finally, he put down his glass. "I think you better explain this in more detail."

She gave him her most beguiling smile. "Let's talk about that later. I'm famished."

"No, let's talk about it now."

"Cody, we don't want to create a scene." She glanced around. "I'll explain it all when we get back to our rooms. Right now, I want to eat."

He glanced out a window. Rain was slanting against the windows.

She gave him a long look, one that women can give a man that is both mysterious and suggestive.

"Shall we have diner?" she asked.

He motioned to the waitress to bring menus.

Cody decided on the New York strip and Erica selected the fried catfish.

During the meal, Erica avoided the subject Cody most wanted to discuss. He had a lot of questions. He wanted answers.

Their conversation became desultory as they ate. When they finished the meal, Cody saw that it was still raining hard. Thunder crashed across the sky.

Erica paid with plastic. Outside they dashed through the downpour to their adjoining rooms.

Before Erica disappeared into her quarters, she shouted from the doorway, "Come on over, but give me a little time, sweetheart."

Cody removed his wet clothes and dropped onto his bed and lay listening to the storm. After a few minutes, a soft rap on the door between their rooms interrupted his thoughts.

"Cody," she called softly, "I've been waiting".

He got up slowly and put on his bathrobe. His cloths were drying in the bathroom. When he went into her room, he found her lying in the bed, a drink in one hand. She was undoubtedly naked beneath the sheet that barely covered her breasts.

"Yours is on the sink, darling," she said, gesturing with her glass at two canisters of drinks she had ordered from the bar.

He got his Old Fashioned and moved toward a chair. "Cody, over here," and patted the bed.

"This is fine," he replied.

She smiled provocatively. "I hate talking across a room."

The infrequent flashes of lightning and the diminished thunder indicated the storm was moving away.

"I can hear you."

"Why are you being difficult, Cody?"

"Am I being difficult?"

She studied him a moment. "All right," she said with disappointment, "if that's the way you want it." She sipped her drink. "I've been thinking about your research, your interest in the wetlands. At Flagler, you can't really carry on your research. You need a better lab, and you need better equipment. Furthermore, you don't need the interruption of teaching."

He tried to anticipate where she was heading.

"I don't get it."

"My proposition is simple. I'm giving you the chance to carry on the kind of research you want. No strings." She paused to let that sink in.

"I'm not sure what you mean by 'no strings.'"

"You won't have distractions, except," she smiled, "perhaps just one, if you want to call it a distraction."

"Please be more explicit."

"I plan to set you up with a lab, a modern, well-equipped one. Isn't that explicit enough."

"The distraction?" he asked. He was now pretty sure what it was.

"Do you like the idea?"

"You're too vague."

"You think I have ulterior motives?" She asked with an injured tone.

"I didn't say that, but I wonder why you would do this."

"Don't you trust me?"

"That's not the point," he replied.

She finished her drink and immodestly threw off the sheet. Cody gazed at her nakedness.

"Look at it this way, Cody, darling. I want you."

He had trouble keeping his eyes off her.

"You would continue to represent the company in its efforts to respect the environment, but you'd do it from here. From your own lab."

"And I'd be nicely out of the way," he retorted harshly. The full realization of what she planned was now obvious.

"Why think of it that way? Think of it as an opportunity to continue your work. Naturally, I'd want to check on how things were going."

Naturally, he thought. "Does Ed know about this?"

"Ed doesn't figure in this arrangement, Cody. And you'll be paid well." A cryptic smile accompanied that last comment.

A lot of guys, he mused would jump at a chance like this. It was the kind of deal men joked and fantasized about. She was offering him sex, along with a lab in which to conduct his research. It was like having the best of all worlds. Sex and research. Or research and sex. Which way would she have it?

He finished his drink and got up.

"It's one hell of an offer," he said, unable to hide his anger over her thinking she could buy him so easily. "And I'll bet," he said icily, "you're great in bed, but I'll pass on that."

"I can have you arrested," she remarked.

He stopped with his hand on the knob of the door adjoining their rooms. "Arrested?"

"If you go through that door."

"What do you mean?"

"I'll scream."

"You'll what?"

"Let's not make a scene. You are in my motel room. How would that look to the police?"

He stared at her.

"I'm ready to give you a damned good time."

He was sure of that.

"You really think you can force me into doing anything you want by offering sex and a research lab? That I'll crawl into bed with you like a thankful puppy, slavering to do your bidding."

He turned the knob. "I'll take my chances with police. But I want you to understand one thing, you are not forcing me to fuck."

With that he slammed the door behind him.

There was only silence.

* * *

They engaged in little conversation on the drive back the following morning.

"You had better think it over, Cody," she remarked, finally. It sounded like a threat.

"What's there to think about?"

"I don't think you appreciate the deal I'm offering you."

"Possibly, but now that I'm fired, what difference does it make?"

She glanced at him. "Who said you were fired. I may think you're a sonofabitch, but you still have the job."

Perhaps she did not intend to give up on him. He thought of the old saying about a woman scorned. Scorned or not, Erica wasn't the type to take rejection easily.

She stopped at the marina entrance.

"We have some more talking to do, Cody," she said as he got out.

He watched her burn rubber down Riberia Street, then he turned to enter the marina. Rocky and Jeanette along with several others were standing outside the office. Rocky broke the news.

"They got Water Rat."

CHAPTER TWENTY-TWO

"What do you mean?" Cody demanded.

"He'd been drinking," Rocky explained, "and went over to the construction site. Apparently no one saw him. The damn fool climbed a tree and said he wouldn't come down until they stopped ruining the wetlands."

"And?"

They tried to talk him down. When that failed, they called Gear."

Rocky mopped his forehead.

"When Gear arrived, he told his men to forget about talking and to get him down one way or the other. Some guy started cutting into the tree, but it was rotten and before he cut very far, the tree went down. They pulled Water Rat out of the tangled and broken branches. A two-inch limb had gone through his chest."

"They killed him?" Cody shouted. "Jesus Christ, what have I done?"

"It wasn't your fault, Professor," Jeanette said.

"I never intended that anyone take things into their own hands."

"He was drunk," Rocky said. "You can't blame yourself."

Cody smashed a fist into his palm.

Rocky put his hand on Cody's shoulder. "You got to understand that what he did had nothing to do with you." He paused. "What you need is a good stiff drink. Let's go down to my boat."

"Listen to Rocky," Jeanette said softly. "Water Rat did this strictly on his own."

Pain etched Cody's face. "Gear isn't going to get away with this."

Jeanette quickly explained that the funeral service would be held at ten in the morning at the Baptist church on Anastasia Island. Mitch had arranged everything with Shawna's help.

For a moment, there was a look of anguished indecision on Cody's face. Then his expression changed. He ran his fingers through his hair. Without another word he strode over to his Harley.

"Professor," Jeanette yelled, "what are you doing?"

Cody sped out the marina gate, and with tormenting visions of Water Rat in the tangled branches of the fallen tree, roared down Riveria Street.

All he could think of was that Water Rat had obviously gotten the crazy idea to confront the developers from what Cody and the others had done.

He swung onto King Street and, weaving through traffic with horns blaring around him, he roared down the Avenita Menendez.

At Gear's office, he jumped the bike onto the sidewalk and put it up on its stand. Without removing his helmet, he raced up the steps, threw open the door, and darted into Gear's building. He caught a glimpse of Erica's startled face as he ran past her office.

As he barged into Ed Gear's office, the owner of Condo Development, Inc. looked up with startled disbelief. "What the hell is this?" Gear shouted.

Cody glared at him. "You sonofabitch."

"What?" Gear asked.

"Cody, please." Erica had rushed in, and now grabbed him arm.

"You're responsible for Water Rat's death," Cody shouted.

"Listen," Gear said, "you'd better get out of my office before I call the police."

"Not before I'm done," Cody responded, shaking off Erica's hand.

"We warned him. You expect me to baby-sit every kook who has an axe to grind?"

"Bullshit."

Gear leaned both hands on his desk. "Well, I'm sorry it happened, but he had no business on the work site."

"And you pass his death off like that?"

"Christ, Matheson, what you expect me to do?"

"You're a goddamn murderer."

"Fuck you." He reached for his phone.

As Gear punched numbers, Cody bent down, grabbed the base of Gear's desk, and in a swift upward movement, heaved it over, Gear leaped out of the way, as it nearly fell on top of him.

"You dumb bastard," he cried, dropping the phone. His blue eyes burned with rage. He yelled for Mike and Sid.

Cordova entered with Phillips behind him. The engineer's dark eyes went from Gear to the over-turned desk to Cody.

"Jesus, what happened," he exclaimed.

"Get him the hell out of here," Gear yelled, pointing a finger at Cody. "Now!"

"No. Wait!" Erica cried.

Cordova moved toward Cody in a fighter's crouch. Cody threw off his helmet and before Cordova could deliver a punch, fired a hard jab to Cordova's mouth. The suddenness of it briefly surprised the man. Blood trickled from his split lip.

"You think you're a fighter?" Cordova snarled, his eyes hard. As he moved in, Cody kicked him in the groin. The vice-president for engineering yelled and fell, doubled up on the floor

"Damn-it Phillips, don't stand there," Gear shouted, as Sidney Philip stood in the doorway.

Cody glared at Gear. "I'm not through with you. I got enough evidence..." He caught himself before saying more, and strode out of the office.

* * *

The following morning was hot and muggy. People from the marina were crowded into the small church. Jeanette sat on one side of Cody and Shawna on the other. Organ music swelled through the building. When the music faded, the minister delivered a brief eulogy.

"The Lord giveth and the Lord taketh. We may not understand, but there is a meaning to His ways. And He surely had His reasons for taking"—he glanced down at his notes—"Fabius Thornton. I understand he was better known as Water Rat." he paused. "Well, no matter, the Lord has taken him into His fold. We mourn his death, but know he has gone to an eternal peace. Amen."

Organ music rose to a crescendo and subsided. The minister requested that everyone stand for a hymn. When the last strains ended, he asked if anyone wished to say a few words about Fabius.

Mitch got up. "He always paid his bills, and we had no trouble from him."

The minister smiled and nodded. Jugs said she knew he worked on a shrimp boat and probably was a good worker. When no one else spoke up, the minister closed the service with a solemn benediction.

On the way out, Shawna asked Cody to come to her place.

CHAPTER TWENTY-THREE

Cody had to tell Shawna about his trip with Erica. He also wondered what action was planned regarding the two reports he'd given to her.

He had ridden to the church with Jeanette, but explained he had something important to do, so wouldn't be riding back with her. She looked disappointed and glanced skeptically at Shawna. "You going with her?"

"Look, Jeanette, this is business. I got some information about an environmental report. It may make a difference in how we go after the Gears."

"I thought maybe we'd have supper together somewhere."

"Let's make it another time. I'm sorry."

Cody sat at the kitchen table and watched a pretty blonde on the small TV report that Eddie, with winds of 55 mph, was continuing its northwesterly movement toward the Bahamas at ten miles an hour.

It was mid-August, the peak of the hurricane season.

Shawna got a couple beers from the refrigerator and joined Cody at the kitchen table.

"I talked with Brad and Mike about our case against the Gears."

"And?"

"He believes we have a good case for fraud, considering the doctoring of the site evaluations. As for Water Rat's death, it may be hard to prove culpability on Gear's part; however, we see no reason we can't introduce it. We have to be careful, though, on how we do."

"How about negligence?"

"That could be a strong point with regard to Water Rat's death." She paused. "Now, I want to hear about your trip."

He poured more beer into his glass and outlined their trip to Florida's west coast. "The bottom line is she wants me out of the way. She thought the lab idea would do it and she figured sex would be the clincher to my accepting." He smiled and sipped his beer. "Things were somewhat strained on the drive back.'

"I can imagine," Shawna said.

"When she dropped me at the marina, I learned about Water Rat."

"I can't say she didn't look damn inviting lying naked on the bed," Cody offered with a roguish grin.

Shawna ignored the remark as she explained, "I checked into the purchase agreement between the city and the woman who sold the property. Her name was Love, Margie Love. She died about three years ago. Part of the agreement did include a provision that there would be no damage to the environment.

"Mike will preside at the grand jury hearing, and of course I'll be there to present our side." We'll push for an indictment based on Gear's failure to follow the conditions of the sale and that he forged Dr. Chapman's signature. As for Water Rat's death, as I said, we'll certainly bring this up and go for negligence. I think I asked you if Chapman would testify at the hearing?"

Cody said he was sure he would.

"I'm going to go over our strategy with both Mike and Brad. Brad may think he's God's gift to women, but he's a good lawyer."

"What do you think of our chances?"

"I'd say fifty-fifty. Most people in this city are not particularly worried about the environment. However, Mike wouldn't request a grand jury if he thought we didn't stand a chance. Although we can't charge Gear with murder, Water Rat's death won't hurt our case."

She paused and looking directly at Cody added, "There's something I should warn you about. If we go to trial, it's possible Gear's lawyers will go after you. You have great credentials, which can hurt the Gear's case, so the lawyers will try to find something to discredit you. Brad and I will do all we can to prepare you."

* * *

Three days later, Cody looked into the faces of the sixteen members of the grand jury. Seven men and nine women. He wondered how each felt about the environment?

Shawna had given him a brief run-down. Five of the women were housewives. One worked in real estate. Two were waitresses and one was a secretary. four of the men were retired, and the other three were local businessmen.

It was a hot, humid late August afternoon, but comfortable in the air-conditioned courtroom. Shawna, Cody, Brad Longstreet, and Mitch sat at a polished wooden table. Brad had dressed immaculately in a dark suit and bright tie. Shawna looked stunning in a two-piece suit that, without over doing it, revealed her fine figure. Cody wore a white shirt open at the collar and dark trousers. Mitch had put on a clean t-shirt and jeans for the occasion.

Ed and Erica Gear, and their attorney sat at another table. Now and then they put their heads together. Gear frowned, while Erica seemed at ease. When she glanced at Cody, her eyes were hard. Ed Gear ignored the prosecution's table.

Mike Reese, seated on the judge's bench, glanced around the court-room, and announced they would begin.

Ed Gear's riveted his eyes on Reese. He obviously didn't like the situation, and he didn't like the protocol. His attorney, although present, could only advise him and could not speak on the issues or to the jury.

Shawna stood up and looked directly at the Gears. "The state's concern focuses on certain improprieties in connection with the construction of a condominium complex south of the city. It is being developed by Condo Developers, Inc. Actually it's within the city limits. The area lies opposite the Cormorant Marina on Riberia Street and extends from Riberia Street to the Intracoastal Waterway."

She glanced at her notes. "We contend that the planned construction will severely damage the ecology of the site. We also contend that in performing the planned construction, the builders," she glanced at the Gears, "are in violation of the original and subsequent purchase agreements. They have circumvented compliance with that agreement in the bill of sale. In addition, there is the matter of a forged signature on an environmental impact statement submitted to the EPA," She paused. "That the Environmental Protection Agency,"

Gear was on his feet. "That's a damn lie."

Reese glared at him, but spoke quietly. "You will have a chance to give your side, Mr. Gear. Please don't interrupt."

Ed Gear immediately conferred with his attorney.

Shawna waited a moment then added, "An individual, Fabious Thornton, otherwise known as Water Rat, was killed on the proposed site. We admit," Shawna continued knowing what Gear's response would be, "that he was trespassing. That is not the issue. What is at issue here is that Mr. Gear," she looked directly at him, "ordered one of his workmen to begin cutting the tree. We also admit that it is possible Mr. Gear did not know that part of the tree trunk was rotten and when the saw cut partway through, the tree toppled. We content that Mr. Gear was negligent in ordering the tree cut as a means to frighten Water Rat—Fabious Thornton —to come down."

Shawna glanced at Cody. "I call Dr. Cody Matheson."

Cody took a chair provided for the witnesses and was sworn in.

"For the benefit of the jury, Dr. Matheson," she said, "please tell us of your education and experience, particularly concerning wetland research."

When Cody finished, Shawna asked if the St. Augustine Record had correctly quoted him in its June 13 issue.

It had.

Shawna passed a copy to Reese. She also had copies for the jurors.

"It is still your professional opinion that the construction planned at the Riberia street complex will do irreparable harm to the immediate environment?"

"There's no question that the ecology would be devastated from a number of causes brought on by the construction. Not the least of which is pollution."

Despite attempts by his attorney to stop him, Gear was on his feet "Our own evaluation shows just the opposite. In addition, we have received permits for construction from the EPA. "I strongly intend to refute the ridiculous allegations. We have been operating within the law and we deny any wrong doing."

"Mr. Gear," Reese admonished, "I warned you once, I don't intend to repeat myself. Do you understand?"

His face a mask of anger, Gear sat down.

Shawna turned to the Gears. "We have two environmental impact reports."

Both Ed and Erica exchanged glances and conferred with their attorney.

"One showed considerable damage would occur to the area if the planned condominiums were built. The other indicated there wouldn't be any."

Several of the jurors looked confused.

One of the women spoke up. "I'm afraid I don't understand." Others nodded.

"I will clarify," Shawna said. She walked over to her desk and retrieved the two reports. She gave one to Reese and additional copies to the jurors.

"Where did you get those reports?" Ed Gear demanded

"They were in the laboratory files," Shawna replied.

Ed Gear's temper was on short tether. "Our files are confidential, goddamnit."

"Mr. Gear, Shawna informed him, "you hired Dr. Matheson as director of the lab. Are you saying he has no right to examine the files?"

"I don't give a damn about his examining them, but I object to those files being aired in this way."

Reese held his temper as he told Gear that the reports were admissible.

"Your honor," Shawna said, addressing Reese, "Dr. Matheson found the reports during the course of familiarizing himself with the laboratory. Mr. Gear doesn't want them aired, because they are detrimental to his defense."

Erica glared at Shawna and then at Cody.

Ed conferred with his attorney, then stood up.

"May I speak on this, your honor?"

"If there's no objection from Miss Gregory, who, I believe, is now finished."

"No objection," Shawna replied.

Gear explained, "Dr. Chapman did not understand the situation when he conducted the initial evaluation of the construction site. He did not know exactly how we planned to go about constructing the condos. We had no intention of doing any harm to the areas ecology."

He paused and glanced at his notes. He was clearly working hard to keep his calm.

"Once this was explained to him," Gear continued, "he revised his original report. It's as simple as that, a case of misunderstanding. We agreed I would sign it. Dr. Chapman gave me the okay to sign his name."

"A power of attorney?" Shawna asked.

"Well, no," Gear replied, "not exactly."

"What, then, exactly?" Shawna pressed.

"An agreement."

Shawna walked over to her table and opened a folder. "We have a statement from Dr. Chapman, who was unable to appear today, that he agreed you would sign your name, with the notation it was for Dr. Chapman. Isn't that correct?"

"Perhaps Dr. Chapman's memory is faulty on this point," Gear remarked.

"Then the question is, Mr. Gear," Shawna continued, "why did you make the signature look like Dr. Chapman's?"

Gear's eyes hardened. "It was simply a matter of expediency."

The jurors were listening closely.

Gear continued. "We have a representative of the EPA who can set the record straight. But before we hear from him, I wish to get back to Water Rat's death. That was purely an accident. And, as Miss Gregory admitted, he was trespassing. We did not know the tree was rotten. I only wanted to frighten him into coming down by making him think we would cut down that tree. I did not intend to have it cut down."

Gear paused, now I would like you to hear from our representative from the EPA."

A short, overweight man took the witness chair, as Cody went back to the table and joined Brad.

"Please tell us your position with the EPA," Reese said.

Clark Smith gave his name and his position as deputy director in charge of issuing building permits. His office was responsible for reviewing requests, investigating them, and rendering decisions based on the investigations and appropriate laws.

"Are you aware," Gear asked standing at his table, "of our environmental impact statement that indicates there would be no harm to the environment with the construction of the planned condominiums?"

He was.

"Did you see any problems with it?" Gear asked.

"None. And because of the report we received we issued the necessary permit." He was obviously justifying his and the EPA's action.

"May I ask a question?" Shawna addressed Reese?

Reese nodded.

"Did you send anyone out to inspect the area?"

Smith didn't answer immediately, then shook his head. "Usually we don't, if we receive a statement from the Flagler laboratory." He glanced at the Gears, then back at Shawna. "One of the functions of the lab is to eliminate the need for our agency to conduct costly environmental investigations. It saves the taxpayers money," he added, with a self-appreciating smile.

Shawna continued. "Another question, Mr. Smith. Are you aware that the original deed drawn up when Mrs. Love sold the land to the city stipulated no harm to the ecology of the area would occur under any circumstances?"

Smith looked confused. "I am not aware of any deed."

"I have copies of that deed," Shawna said, and handed one to Reese.

"We need to get back to the matter of a forgery," Shawna said. "Did you know that Dr. Chapman's signature had been forged?"

"We had no reason to suspect that," Smith replied.

Gear and his attorney put their heads together. Gear asked, "If Chapman gave permission for us to sign his name, what is wrong with that?"

"Miss Gregory?" Reese asked.

"Dr. Chapman denies giving you permission to sign his name. We have his statement to that effect. It is our contention that not only did the forged signature influence the EPA's decision, but that the signature was purposely made to look like Dr. Chapman's. It was unlikely the EPA would question Dr. Chapman's signature, whereas they would question the report if it was signed by Mr. Gear, since, of course, he is the developer."

"Did you examine Dr. Chapman's signature?" Reese asked Smith.

"Examine it?" Smith replied. "We assumed it was Dr. Chapman's." He paused. "You see, we are very busy. We have a lot of permits to consider. We simply can't check out every signature on a report."

"Please understand, Mr. Smith," Shawna continued, "you are not the subject of the grand jury's investigation, nor is the EPA. The point I am making is that you were led to believe the report on the environment was properly prepared and signed by Dr. Chapman, who was the laboratory director. Your understanding was that Dr. Chapman was responsible for its contents. And since the contents showed no harm to the environment, your office decided to grant a permit."

"That's a fair estimate," Smith replied.

Shawna faced the members of the grand jury. "Our position is that Mr. Gear had no intention of altering his original plans, which, when carried out, would do irreparable harm to the environment. We also believe Mr. Gear misled the environmental protection agency into believing that the report they received was a true and valid one based on Dr. Chapman's finding."

Both the Gears were on their feet. "That is blatantly misconstruing the truth," Ed Gear shouted."

Their attorney struggled to restrain them.

"Everything was done according to law," Erica said loudly.

"Absolutely," Ed Gear shouted.

Reese gaveled for quiet.

"She's distorting the truth," Gear cried. "I refute the ridiculous allegations."

Both Gears were yelling at once.

Again Reese pounded his gavel.

Shawna addressed Reese. "It seems there is a pattern to defraud here. The Gears did not adhere to the original desires expressed in the deed, when Mrs. Love sold the land. The Gears were fraudulent in using Dr. Chapman's signature on a report knowing it did not represent the true situation if condos were built. And" she paused "Mr. Gear was negligent in his handing of the Fabious Thornton death. However, we believe beyond any doubt that to build condominiums on the wetland site would drastically harm the environment. We intend to present another witness, Mr. Mitch Larson, who is also an authority on the ecology and the environment, to corroborate Dr. Cody's statements. We contend that Mr. Gear's argument of no environmental harm is based solely on a report prepared under false circumstances. We contend that Dr.

Chapman was threatened, in so many words, that he would be fired as director of the lab if he did not prepare a second report."

Gear raised his fist. "There is not a modicum of truth to that."

"I submit," Shawna said looking directly at the grand jury, "we can't substantiate that Dr. Chapman was actually threatened, that is, beyond a question of doubt, but we can show that environmental facts have been misrepresented and that, as I said before, Mr. Gear is guilty of forgery and misleading a state government agency. Let alone," she added, "negligence in Fabious Thornton's death."

Reese addressed Gear. "Do you have anything else you wish to say at this point, Mr. Gear? Do you have additional witnesses to back up your contentions?"

The president of Condo Developer conferred briefly with his attorney. "We cannot overstate that we are being maliciously attacked and that once our position is explained, it will be perfectly clear that there are no grounds for the accusations." He paused. "Should this case go to trial, you can be assured we will present evidence that will clearly show we are innocent of the charges. However, we are not at the moment prepared to counter these malicious lies. We need more time to prepare."

"You were made aware of the charges," Reese said, "and you had time to develop a response to them. You realize you have the right to present your side at this hearing. You do not have to wait for a trial, if there is to be one. Do you understand your position? You are free to call other witnesses."

The Gears conferred briefly, then Ed Gear replied, "We will address the charges at the next session."

"I'm sorry," Reese replied, "but unless you can show extenuating circumstances, we will wind this up today."

Erica Gear jumped up again, her eyes blazing. "We are being railroaded. We object that our lawyer cannot address the court, which I think is absurd. How can we counter the accusations without time to prepare? We resent the manner in which this hearing is being conducted."

Reese stared at her. "We are conducting this hearing in accordance with the law, Mrs. Gear. Your attorney should have briefed you as to how we would proceed. I regret you feel the way you do, but I have no alternative at this time but to turn the information presented here to the jurors. Now Miss Gregory," Reese asked, "do you wish to call another witness?'

Shawna turned to Reese. "In view of Mr. and Mrs. Gear's failure to rebut our claims of negligence other than their word against ours—and the documents I have submitted—no, I don't believe it necessary to call another witness."

"Thank you Miss Gregory."

Reese addressed the Gears. "Have you further information you wish to present?"

Both glumly shook their heads

The district attorney turned to the jurors. "Do you have any questions?"

All shook their heads.

"Your responsibility, Reese continued, "is to determine if there is sufficient evidence for the state to proceed with a jury trial. You will deliberate among yourselves. You will not discuss this case with anyone other than yourselves. When you have reached a decision you will notify my office immediately. Are there any questions?"

There were none.

On the way out of the courthouse, Cody turned to Shawna. "What do you think?"

"I could use a drink."

CHAPTER TWENTY-FOUR

The Grand Jury returned its indictment within two days. A trial date was set for the second week in September.

"Fortunately," Shawna told Cody over dinner at a *The Back Door*, a St. Augustine restaurant, "we have Judge Klein presiding. That's a break for us. Although we have a good case, I wouldn't relish being up against Henderson."

As they enjoyed an after-dinner drink, Shawna reviewed the strategy she and Brad had agreed upon.

"Brad and I will work together on this. As of now, we plan to use you as a key witness. We'll call Dr. Chapman and we'll probably want Mitch to testify. We may want testimony from the environmental Protection Agency with regard to their approval of the condo construction. We don't have the list of witnesses the Gear's attorneys will call, but it will be provided. I suspect they'll dig up authorities that will claim no harm will come to the environment with the building of the condos. Despite your credentials, Cody, I'm sure there'll be witnesses that may not agree with your assessments. "We plan to emphasize that the Gears did not honor the original contract when the land was sold. Another strong point in our favor is that Ed Gear is probably guilty of forgery."

Cody ran his hand through his hair. "Mitch can testify as to the pressure the Gears put on him to sell the marina. Even though he didn't finish graduate school, he knows a lot about the wetlands. And certainly he'd corroborate anything I said."

"We have to be careful," Shawna cautioned, "because we can't prove Gear was behind the break-in or any of the so-called accidents at the marina. Same with the beating. However, we can plant some suspicions."

Cody mentioned that a number of potential witnesses at the marina had left. Several days earlier, he had helped the Milsaps cast off. The MacIntoshes had left, and so had Marion Talbert, Jules Rodreques, and Gus Johnston and his wife. It seemed to him, watching them motor down the Sebastian River, that something of himself was leaving, too.

"That won't present a problem," Shawna replied. "With you and Mitch, I think we're okay."

* * *

The following day Shawna phoned Cody with the bad news.

"We just got word, Klein isn't to try the case. Which means Henderson will. Seems Klein's daughter took a group of young people to the condo site to protest. She got knocked down and cut. We don't know the details. You can imagine Klein's reaction—the apple of his eye and all—so he claims he couldn't be impartial conducting the trial."

Aboard the Columbia, Cody watched the latest weather report. Tropical storm Eddie was continuing its northwest track and was moving into the central Bahamas. The winds had increased to 60 miles an hour. No change in direction was forecast over the next twenty-four hours. Interests along the Florida east coast should be alerted to strong winds, high tides, and coastal flooding.

Cody mixed a bourbon and water and lay on his bunk. He reflected on the up-coming trial. He knew very little about trials and the law, but he was concerned whether or not Shawna and Brad had a strong enough case to prevent the Gears from building their condos. He had kept Mitch informed on how things were going; having informed him he might be called to testify.

"No sweat. I look forward to doing anything I can to stop the Gears from hounding me and building those goddamn condos."

* * *

Brad Longstreet approached the jury box.

The trial date had been moved up from the third to the first week in September. Judge Henderson planned to be on a cruise ship to the western Caribbean by the end of the month. He wasn't about to let the trial interfere.

Two attorneys, Jerry Coles and Ralph Cobb sat with Ed and Erica Gear. Coles was tall and thin. Cobb was short and overweight. Shawna was now alone, as Brad spoke to the jurors.

Cody and Mitch sat directly behind her in the row of spectator seats.

"Ladies and gentlemen, you have a very important responsibility. A responsibility to prevent rape." Several jurors exchanged glances.

"I'm referring," he continued, "to the rape of our environment. You are to decide whether the developers will have free rein to ravage our wetlands for a few condominiums.

"In brief, that is what this trial is about. The preservation of our natural heritage. It's about the possibility of fraud and negligence in the death of Fabious Thornton. It's about saving our irreplaceable wetlands, which are in danger of being raped into extinction. "

Brad looked disapprovingly at the Gears and their two attorneys. "The construction of these condominiums on Riberia Street will destroy one of natures most vital ecological wonders."

The Gears' attorneys frowned slightly as they scribbled notes. Judge Henderson peered over his thick-rimmed glasses as Brad continued.

"The site where the construction is planned consists of a wetland interspersed with various indigenous trees. The area is ecologically fragile, which means that it is extremely susceptible to harm from intrusions.

"The site is a natural habitat. The many creatures that live in that wetland have a right to survive. But it is important to know that their survival is linked to natural systems beyond the confines of the wetland area itself.

"The nutrients manufactured there by microscopic organisms are carried out to sea where they provide essential food for fish. These are important to the commercial and sport fishing industry. Not to mention migratory birds that use the area as a feeding ground and resting place.

"It is not a case of the wetlands being isolated entities, swamps, as some people call them, emitting foul vapors, ugly places of black ooze and grass. To the contrary. The wetlands affect areas adjacent to them as well as those far removed. It is like the ripples on a pond when a stone is dropped into it. They expand out in ever increasing circles affecting what they come in contact with."

Brad paused. "Let me reemphasize that the health of the wetlands is vital to our ecology, as well as to health and well-being of migratory birds, to animals, to all the creatures that live in and visit the wetlands.

"These are some of the facts you must weigh in deciding if the continued rape of the wetlands is to be permitted by those who would desecrate one of nature's most beautiful and vital systems.

"But you need not take my word for it. Experts will testify to what I'm telling you, men who have spent their lives studying wetland ecology, experts dedicated to preserving one of nature's marvels.

"Ladies and gentlemen, you must look into your hearts to determine if the devastation of your wetlands is to continue."

He stood for a moment, looking at each juror.

Judge Henderson frowned, scribbled on a sheet of paper, then looked at the lawyers at the defendant's table. "The defense may now make its opening statements."

Jerry Coles got up slowly, frowning. A thin man, he stood slightly bent, hands in his pockets.

Finally, he half turned and gestured toward Shawna and Brad. "Well, that was an interesting bit of information about our wetlands." His manner was that of a country boy duly impressed by a learned authority.

"'Course I don't claim to know all about wetlands, but I like that bit about ripples in a pond. Reminds me of an old pond I swam in as kid." He allowed himself a few moments reminiscence. "Now one thing I remember about those ripples," he paused smiling, "is when they reached the edges of the pond they really didn't do much." He grinned coyly.

"Which gets me to thinking how much impact those wetlands have—if they're like ripples, as Mr. Longstreet suggested."

Several jurors smiled.

"Mark one up for the defense," Brad muttered unhappily.

Coles glanced down at his feet then at the jurors "You know, there was one item Mr. Longstreet didn't mention. He never mentioned people.

"Now I don't say those little creatures living in grass and mud aren't important. No sir. But I say it strikes me strange not to mention people."

He spread his palms on the front of the jury box and looked at the jurors with raised eyebrows.

"The way I see it, aren't people important, too? People need places to live. They need recreation areas. Are we going to deprive these people of their well-earned and deserved life styles while we protect a bunch of lugworms, barnacles, clams, and what not squirming around in the black mud?" He raised a hand. "Now don't misunderstand me. I think those things have a right to live. Yes sir. But are we going to let all kinds of creatures deprive us of our rightful place in the scheme of things?"

He loosened his tie. "Now I like to fish. And I suspect some of you do, too. And I think we all agree that we don't want to do anything to hurt our commercial fishermen. Those guys work hard and long to bring food to our tables. And they deserve our admiration and consideration. However, I just can't fathom how putting up a few condos down there off Riberia Street is going to put those nice people out of work or hurt my chances of doing' a little fishin'. I just don't see how that will happen.

"'Course I'm not as learned as those fellas' who'll be called to testify about the wetlands. Maybe I don't know all about Spartina grass, lugworms, and all. But I do know about people. I do know we just can't ride slipshod over our fellow men, and women, just to keep a few crabs from multiplying.

"Now we have our witnesses to call, of course. And you'll be interested in what they have to say. Let me conclude that the way we see it, the building of a few condos isn't going to harm the wetlands. My clients," he turned to include them with a wave of his hand, "don't plan to harm the ecology of the area. They plan to take every precaution to protect all the little creatures living in the area. Fact is, they plan to turn the area into a nature preserve. People will be able to take pleasant walks there and enjoy getting back to nature. That hardly seems to me like they're planning on devastating, as my colleague claims, this area."

He paused as though thinking of something more to say. "Well I guess I've taken up enough of your time."

With a smile to the jurors, he returned to the defendant's table.

Judge Henderson looked at Shawna and Brad. "Are you ready to call your first witness?"

Brad stood. "Yes, Your Honor. I call Dr. Ron Chapman."

The one-time director of the Flagler Environmental Laboratory looked uneasy as he took the stand. His face worked and his eyes darted about the jury room. When he was sworn in, Brad approached him, smiling. Chapman removed his glasses and cleaned them with his handkerchief.

"With regard to the development on Riberia Street, did you conduct an environmental impact statement?"

Chapman explained he had, and had given the report to Mr. Gear.

"What was the finding of that study, Dr. Chapman?"

Chapman glanced around. He blinked. "I recommended that construction would do great harm to the ecology of the site."

"What was Mr. Gear's reaction?"

Chapman ran his hand across his forehead. "He said I didn't understand the true situation."

"What did Mr. Gear mean by 'true situation?'"

Cobb raised a pudgy hand. "Objection. It calls for the witness to speculate."

"Sustained," Henderson intoned.

"Did Mr. Gear explain what the 'true situation' was?"

"Yes sir. He said that since he would comply with the conditions of the sale of the land no harm would come to the environment. But I explained that I didn't see how he could avoid hurting the environment. I told him I couldn't alter the facts that I'd put in the original impact statement."

"And?"

"Mr. Gear said I'd better redo the report."

"Did you?"

Chapman swallowed. "Well, yes."

"Even though you knew it would be incorrect?"

"I didn't want to jeopardize my retirement."

"Did Mr. Gear threaten that?"

"Objection, Your Honor."

"Sustained."

"What happened next?"

"Well, I rewrote the report, but refused to sign it. Mr. Gear said in that case he would."

"And did he?"

"Well, yes and no."

"Would you please explain."

"He signed my name."

"He forged your signature?"

"Objection," Coles was on his feet. "This calls for a legal opinion by the witness, Your Honor."

"Sustained."

"Doctor, did your name appear on the report?"

"Yes sir."

"Did you sign the report?"

"No."

"Did you object to your name being put on it?"

"Yes," Chapman answered, his head bobbing and his lips working. "But Mr. Gear ignored my protests."

"And that report was submitted to the DER, is that correct?'

"Yes."

Brad turned to the judge. He had no further questions at this time.

Henderson nodded. "You wish to cross examine, Mr. Coles?"

Coles got up slowly and shuffled toward the witness stand. He stood for a moment as though collecting his thoughts.

"Well now, I just got a few concerns that maybe you can help me with, Doctor. I assume that the DER thought you signed that report, is that a reasonable assumption?"

Brad spoke up. "Objection, it calls for speculation by the witness."

"Overruled," Henderson said.

"Christ," Brad muttered. Shawna laid her hand on his.

"I guess that's what they thought," Chapman answered.

"You knew the DER approved a report you had not personally signed."

"Yes sir."

"What did you do about it?"

Chapman glanced around. "Nothing."

"You let a report go through with your signature, but which you didn't sign and did nothing?" Coles tone was accusatory.

Chapman wet his lips. "I didn't think there was anything I could do, under the circumstances."

"Under what circumstances?" Coles pressed.

"Objection, Your Honor, he's badgering the witness."

Henderson glared at Brad Longstreet. "Overruled."

"Well, now," Coles continued in a belligerent manner, "even though you claim you didn't sign the second, you did nothing about it. To me, Doctor, that strongly suggests at least tacit approval. Am I correct in that?"

"Objection," Brad exclaimed. "He's leading the witness."

"Overruled."

"No further questions," Coles said. He wore a self-satisfied smile as he shuffled back to his table.

Judge Henderson banged his gavel. "We'll recess for lunch and convene at two o'clock this afternoon." He admonished the jurors to talk to no one about the proceedings, emphasizing that if they did he would have to sequester them.

CHAPTER TWENTY-FIVE

Cody, Shawna, and Brad, had lunch in a small cafe on San Marco Boulevard.

Brad glared at his glass of beer. "That sonofabitch Henderson will lose this case for us. He didn't grant one of my objections."

"The trial isn't over," Shawna remarked.

"That may be," Brad retorted, "but we're off to a damned lousy start. I think Coles has the jury eating out of his hand."

"We better put Cody on the stand so we can make some badly-needed points with the jury."

Brad agreed.

"Coles surprised me," Shawna remarked, "when he didn't ask for a dismissal on the grounds we haven't proved a crime or felony."

"They may figure," Brad replied, "they got an open and closed case. But in that case why continue?"

"To make a stronger statement for the developers," Shawna remarked,

"I'm sure at this point," Shawna said, looking at Cody, "that they figure you're the biggest threat to their case. You're an expert in your field. They're going to look for a way to discredit you. And they may also be leery of Mitch,"

They finished their meals, paid, and returned to the courthouse.

With the resumption of proceedings, Shawna called Cody to the stand. After he was sworn in and had given a rundown on his education and experience, Shawna asked, "Did you conduct a survey of the Riberia construction site?

176

Cody replied he had.

"Would you tell us what you found?"

"Its ecology is very fragile. By that I mean any intrusion would upset its delicate balance."

"Would you explain that please, Doctor?"

"Typical of most wetlands, this one formed over several hundred years. During this time, a special type of grass, called Spartina, became established that had to adapt to a harsh salt-water environment. Also a variety of marine and animal life, as well as insects and spiders, learned to survive in less than optimum conditions."

Shawna smiled. "It sounds like a veritable zoo."

Several jurors smiled.

"However," she continued, "wouldn't it be possible to establish a workable relationship between the proposed condominiums and the variety of species that inhabit or use the wetlands?"

"No."

"Why not, Doctor?"

"The most obvious reason is that any construction would destroy the habitats of the important species indigenous to the area. Also, pollution would make it impossible for various forms of life to continue to exist. Such pollution would come from roads, sewerage, insecticides, and pesticides.

"However, we are overlooking an important facet of the wetlands."

Shawna did not immediately ask what that was, but let a moment pass to create suspense for the jurors.

"What is that?"

"The wetlands provide both food and a habitat for a variety of fish, animals, and birds. For example, mullet and menhaden, which are important to the commercial fishing industry, enter the wetlands shallow waters to spawn. Shellfish, such as oysters, crabs, and shrimp feed directly on the detritus and algae produced by the wetlands. Many small fish that live in the wetlands and its estuaries become food for commercially valuable fish."

"You're saying that our wetlands are invaluable for both sport and commercial fishing?" Shawna asked.

"Absolutely. But they can also serve as excellent laboratories. Research shows that shallow wetland ponds can produce two-hundred and fifty to four hundred pounds of fish per acre, and that some one hundred pounds of crabs and three hundred pounds of shrimp can develop in wetlands areas without cultivation. Wetlands can also be used to study how plants, animals, and fish adapt to adverse and changing environmental conditions."

The Gear's attorneys were scribbling furiously.

Coles finished writing and stood up. "I object, Your Honor. Where is all of this leading?"

Henderson turned to Shawna. "I presume there is a point to this lecture on the wetlands?"

"Yes, Your Honor. I am trying to show how vitally important the wetlands are."

Henderson sighed. "All right, continue, but I suggest you get to the point."

Shawna told Cody to continue.

"The wetlands are also valuable in controlling flooding from the sea. They take the brunt of powerful waves, causing them to break before they can do damage to structures near the coast. If dredging and building homes and shopping centers destroy our wetlands, these in turn might well be destroyed by storms."

Coles rose unhurriedly. "I object. We're not talking about wetlands adjacent to the ocean,"

Henderson nodded. "You have a point counselor; however, I would like to hear what Dr. Matheson has to say."

"Thank you, Your Honor," Shawna said.

Cody continued. "The Riberia Street site is invaluable in controlling flooding. By building of a channel for boats to dock at the condo piers and by thinning the trees would greatly increase the chance of inundation. It is very possible that high water associated with tropical storms, given the right wind conditions, could sweep into the site and cause considerable property damage. In other words, the wetlands provide protection for coastal areas from flood and storm damage. They are also aesthetically valuable as they provide variety on our highly developed coastline."

"I suppose," Shawna remarked, "that one wetland is just like any other."

"Not at all. We must not make the mistake of thinking so or thinking they never change. But they do. As Dr. Robert Weeden, an authority on natural resources management, claims, there is a great deal of diversity among wetlands in different areas. They may appear to look the same, but it would truly be unique if one wetland was actually the same as any other.

"The wetlands, where construction is planned are very much different from others. We cannot regard it as just another to be treated the same as others. There's also the aspect of change. We think much of what we see around us changes very little. This is not so. The wetlands are constantly changing. Changes in water levels and salinity, warm and cold spells—all affect the great diversity of life within the wetland.

"We don't know exactly what effects a parking lot or roads will have on the life within the wetlands. How will these affect the growth of clams, oysters, or the supply of fish for the various fish-eating birds, we aren't sure, but we can assume it certainly won't be beneficial. We do know that changing the ecology of the wetland will adversely affect both sport and commercial fishing."

"In other words, Doctor, we shouldn't tamper with nature without first knowing the consequences."

"Especially in view of the delicate ecological balance of the wetlands, as I have already mentioned."

Coles rose when Shawna finished and approached Cody. He stood for a moment without speaking.

"Tell me, Doctor Matheson, why did you leave your last position with the Virginia Polytechnic University."

Cody was surprised at the question. "I wasn't happy with my job."

"Isn't it true you left because you didn't get tenure?"

"Not exactly."

"What do you mean by 'not exactly.'"

Matheson explained the terms of his hiring, that the department head had promised him he'd head up an environmental laboratory and be given tenure after the first year. He added, "It became increasingly clear that there would be no environmental laboratory."

"Why didn't you get tenure, Doctor?"

"It was a matter of publish or perish, I suppose."

Coles smiled knowingly. "I suppose, you failed to publish and subsequently perished?"

Brad objected. "He's taunting the witness, Your Honor."

"Over ruled. Answer the question," Henderson advised.

Cody explained he did publish, quite a bit, but did not say in what type of publications.

"What about your research?"

"There was no opportunity to do the kind I felt important."

Coles nodded. "I see. In so many words, then, you had to leave because you failed to be tenured and publish as required by the university?"

It was obvious he was trying to create a question about Cody's integrity and expertise.

"Objection," Brad called. "He'd putting words in the witness's mouth."

"Overruled."

"I resigned because I was decidedly unhappy with the work I was doing and that the university reneged on initial promises."

"Reneged?" Coles asked. "How so?"

"I was promised a research facility. It never developed."

"Did your wife want to leave?"

Cody stared at Coles. "I don't understand your question. Furthermore, I don't think it relevant."

Coles noted the look on Cody's face and realized he might be on to something.

"You don't think it relevant? That's not for you to decide, Doctor. As you know, wives often play an important role in a professional's career. Was she upset over your leaving the university?"

He couldn't know, Cody thought, about Karen's antics or her escapade with Raskin.

"I don't know, because we were in the process of a divorce."

Coles studied Matheson carefully. "A divorce?"

"That's right."

"Would you to tell us what precipitated that?"

Brad was on his feet. "Your Honor, I object. There is no relevance between Doctor Matheson's divorce and this case."

Henderson realized he couldn't overrule this so easily. He frowned at Coles. "I will have to sustain that objection."

"Very well," Coles said. "But I would like to know what role, if any, the former Mrs. Matheson played in the witness's decision to leave the university."

Once more Brad objected. "Again relevancy, Your Honor."

"I'm sorry," Coles said, before Henderson could rule. "I withdraw the statement."

He felt he'd gone as far as he could for the time being, but he also felt he had pried into something that might bear investigation.

"I have no further questions, Your Honor."

When Cody resumed his seat, Brad called Mitch Larson to the stand. Mitch gave a brief account of his education and experience. He explained he left graduate school, prior to earning his doctorate.

"But you majored in ecological studies, isn't that correct?"

"I completed all my studies, but not the dissertation."

"Then you, are what is called ABD—all but dissertation. Am I right on this?"

"Yes."

"I gather, then, that you have had a considerable educational background in the ecology."

"That's right."

Coles was on his feet. "I object, Your Honor. We've heard enough about Mr. Larson's background. He doesn't have a PhD and, therefore, his testimony may be suspect."

"Sustained. Mr. Longstreet, let's get on with what Mr. Larson has to say."

"Yes, Your Honor."

Brad addressed Mitch. "We've been discussing the wetlands on Riberia Street where the proposed construction would take place. Do you agree with Doctor Matheson, based on your education and experience," he glanced at Gear's attorneys and Judge Henderson, "that construction of the condominiums would be harmful to those wetlands?"

"Definitely."

"I understand the Condo Developers, Incorporated, are interested in obtaining your marina."

"They have made several offers."

"And you weren't interested?"

"I don't wish to sell. And their threats only added to my resolve not to."

"Did you say, threats, Mr. Larson."

"I object, Your Honor," Coles said. "What's this got to do with the construction on the wetlands?"

Henderson stared at Brad. "I'm not sure myself just where this is leading, but I'm going to allow you to continue, since I would like to hear about these so-called threats."

"Please elaborate," Brad said, "on what you referred to as threats."

"On several occasions I was told I would regret not selling. And then things began to happen."

"Who made the threats?"

"I object," Coles was on his feet.

Henderson held up a hand. "I would like to hear the answer, Mr. Larson."

"Juan Cordova and Sidney Phillips."

Everyone's attention was riveted on Mitch.

"Who are these people?" Brad asked.

"Cordover is an engineer for Condo Developers and Phillips works in their real estate area."

"And they threatened you?" Brad persisted. "How?"

"They said things might not go too well for me, if I didn't agree to sell."

"What did they mean by 'things might no go too well?'"

"Objection." Coles cried out. "This calls for speculation by the witness."

Judge Henderson frowned. "I would like to hear about these threats. Overruled."

"They didn't say just what would happen, but after the threats, a number of accidents occurred at the marina. For instance, a cable broke. A stand holding a boat on dry dock was cut almost through. Various items were stolen, or at least they disappeared and could not be found.'

"You don't know for sure if Cordova or Phillips were responsible, do you."

Brad had planted in the jurors' minds the idea that the accidents might be caused by Condo Developers, so he risked asking a question Mitch could not answer, knowing full well, the Gear's attorneys would object, if he answered in the affirmative..

"No, of course not, but since the things that happened followed on the heals of the threats, there seems to be a connection."

Coles leaped up. "I must object, Your Honor. This is purely speculative. There is no proof of the allegations."

Judge Henderson deliberated a moment. "I agree at this point. Sustained." He advised the jury to disregard the witness's last remarks.

Brad turned back to Mitch Larson, but for a moment said nothing, creating what he hoped would be a momentary suspense. Then he said, "You have, I believe, something to say regarding the wetlands. Something we haven't touched upon."

Coles and Cobb exchanged glances.

"What I am about to say may sound, to some, a bit bizarre; however, many eminent scientists subscribe to the theory."

Mitch shifted his position, as though girding himself for what he was about to relate.

"You see we are accustomed to thinking everything around is separate. That is only natural, because we have to live in a material world. Our senses create this illusion out of necessity. Behind this illusion is an interconnectedness that pervades the entire universe."

"Now hold on," Coles got up slowly. "Your honor, what has this to do with the case at hand?"

Henderson glared at Mitch. "I'm inclined to agree. I don't see the relevancy."

"Your honor," Brad replied, "if I many continue the relevancy will be made apparent, I assure you."

With a look of reluctance, Henderson told Brad to carry on, but he had better show how this testimony relates to the wetlands."

"Thank you," Brad said, and turned to Mitch. "Can you explain this interconnectedness as it relates to the wetlands?"

"Certainly. One area of my studies at the university dealt with this phenomenon. The interconnectedness found throughout the universe has been demonstrated at the subatomic level of matter through studies in quantum physics. It is also manifested in the macroscopic or larger world around us. However, we just don't see it in our daily lives.

"The highly-respected philosopher Alfred North Whitehead argued that beneath the reality of the universe is an underlying web of connections. But, of course, in order to survive, we have to focus on one thing at a time. As a result, we have learned to screen out the connections."

"You're implying," Brad said, "that this connectedness applies to the wetlands."

"The point is that a great deal of irreparable harm will occur to our wetlands by the construction of the condominiums, roads, parking area, tennis courts, and shopping areas. The wetlands may become uninhabitable.

"However," he paused to emphasize his next remark. "The damage would not be restricted to these wetlands."

The jurors listened intently.

"Through the interconnections I have mentioned, there is every reason to believe that other wetlands will also suffer. Like someone with a cold spreading germs."

Cobb was on his feet. "Your honor, this is ridiculous. It defies all reason."

Henderson frowned. "I wish to hear more of this," he said, and nodded for Brad to continue.

"In brief then," Brad said, "the destruction of these wetlands will cause the destruction of other wetlands. Something like sympathy pains people feel."

"Yes, but there's another point," Mitch said, "I must emphasize we cannot look at our wetlands and the life within them in isolation. It is very possible that the interconnectedness I mentioned involves what the scientist Rupert Sheldrake calls morphogenetic fields. This is a new kind of field, a force that connects each individual with all others in its species. Thus each species has a group mind, as Sheldrake calls it."

"For heaven's sake," Coles interjected, rising. "I've never heard such balderdash. Must we listen to this, Your Honor?'

Judge Henderson sighed. He stared for a moment at Coles. "I'm going to let him pursue this, for the moment."

Mitch sat back, aware of the amazed expressions on everyone's face.

"I certainly understand this may sound strange. But a lot of strange ideas are emerging today. There has been considerable research that indicates Sheldrake is right. What happens to one wetland may well happen to other wetlands through their morphic fields. The wetlands are part of a much larger ecological pattern, as Doctor Matheson alluded to."

Shawna glanced at Matheson.

"The crabs eaten by birds lift new avian wings each spring toward the northern reaches of the birds' migration. The detritus that drifts from the wetlands estuaries grows scales on commercial and sport fishes, and the Spartina stalks support multitudes of microorganisms essential to the complex food chain of the wetlands. In countless ways our wetlands shed their illusory isolation and reveal an essential symbiotic relationship far beyond their boarders."

Brad waited for a moment. The courtroom was silent. Shawna stared at Mitch. Although they had all discussed whether they should allow Mitch to testify about this matter, they had agreed to see what would happen, It was, they had admitted, a gamble, but one they agreed to with reservations.

Shawna noted the intense interest on the jurors' faces, as well as the perplexed expression registered by Cobb and Coles. Both Ed and Erica Gear stared at Mitch with dark frowns of incomprehension. It appeared they had no idea whether or not what Mitch revealed hurt or damaged their case. But there was something in their expressions that sparked a warning in Shanwa's mind.

Brad waited, letting the silence in the courtroom create a backdrop to of profundity to Mitch's words.

Slowly, he turned to Judge Henderson. "No further questions, Your Honor."

Henderson spoke to the defendant's attorneys. "Do you wish to cross-examine?"

As Coles glanced up, his face a study of troubled concern. "Not at this time, Your Honor."

"In that case," Henderson said, banging his gavel, "we recess until ten o'clock tomorrow morning."

CHAPTER TWENTY-SIX

"You may call your first witness, councilor," Judge Henderson said.

It was ten o'clock the following day. Cody had spent a sleepless night thinking about the trial and listening to the rain. Then rising, he glanced out one of the Columbia's portholes. Low dark clouds scudded across the sky.

Before dressing, he turned the TV on to a local channel. The winds were now twenty miles an hour with higher gusts, the announcer declared, and reported that Eddie was moving through the Bahamas, continuing its northwesterly movement. Interests along Florida's east and northeast coast should continue on the alert for high winds heavy surf, and coastal flooding.

Now inside the courthouse, Cody once more listened to the rain beating against the windows as Coles stood up.

"We call Dr. Edgar Carter."

Carter, a small, round, middle-aged man took the stand. He folded his pudgy hands over his stomach. As he glanced at the jurors, he smiled confidently, explaining, in response to Coles' question about his background, that he'd been in academia some twelve years, the last five with the Florida State University. He had earned his doctorate in agronomy at Syracuse University. Since coming to Florida State he had done considerable consulting in the area of environmental concerns.

"You are," Coles began slowly, smiling as he rocked back on his heals and smiled, "familiar with the Riberia Street wetlands, are you not?"

"Certainly," Carter said.

"Now, we've heard testimony that construction at this site would harm the site's ecology. Seems the plaintiffs kind of figure those little clams, worms, and other creatures are in danger of being annihilated. Do you agree, Doctor?"

Without hesitation Carter said he did not. He exhibited the archetypical manner of a university savant. "It is my professional opinion that concerns over the wetlands are overblown."

"Would you explain, Professor."

"Of course." He replied as though addressing students. "The various species that inhabit the wetlands are, by nature, rather hardy. This means they can readily make a comeback when their habitat is disturbed. The idea that the clams, oysters, and other creatures could not live in the disturbed area is, frankly, nonsense." He sat back obviously satisfied with his pronouncement.

Where, Cody wondered, was this guy coming from?

"Then you see no reason not to construct condos at the site, provided, of course, care is taken to disturb the natural environment as little as possible.

"That is my opinion."

"Well, that's interesting, Doctor." Coles smiled. "We sure don't want to see any of those little creatures harmed. But I'm a little concerned, because witnesses for the plaintiff claim building condos will harm sport and commercial fishing."

Carter shifted in his seat and crossed his legs. "That is bit far-fetched. There'll be plenty of room for sustained survival after the condos are put up. As for increasing the potential for flooding by cutting down a few trees and putting in parking lots and roads, well, those who think that have not done their homework."

Cody and Shawna exchanged glances.

"You like to hunt, Dr. Carter?" Coles asked.

"Why yes."

"So do I, A little fresh air is good for the soul, wouldn't you say? But what I'm getting at is this: would putting up those condos discourage those migratory birds from stopping off to get something to eat and rest?"

"Like I said," Carter replied testily, "the construction isn't going to cover the entire wetland tract. Therefore, it's obvious to anyone that there'll be plenty of space for migratory fowl."

Coles smiled as he turned to face the jury. "Now I find that pretty darn interesting and, of course, reassuring. A little different from what we've been hearing. It seems the oppositions been using scare tactics."

"Objection," Brad said. "There are no grounds for that statement."

"Overruled," Henderson replied.

Coles continued, now facing the jurors. "I'd say that Dr. Carter has addressed our concerns very nicely. No further questions."

Henderson looked down at Shawna and Brad. "Do you wish to cross-examine?'

Brad said he did.

"You testified," Brad began, "that you did considerable consulting. Was that for the university?"

Carter returned Brad's gaze. "In my position I am free to do private consulting, although it would not be unusual for the university to request my services as a consultant."

"Has the university done so?"

Carter scowled. "Well, the situation hasn't developed that would require my expertise."

"I ask, Doctor, has the university availed itself of your consulting services."

"No.'

"Then all your consulting has been done privately. Is that correct?"

"I have substantial credentials that..."

"Please answer," Brad cut in.

"Your Honor," Coles spoke up, "he's taking an unduly aggressive role with the witness."

Henderson glared at Brad. "Be careful, councilor."

"Yes, Your Honor." He turned back to the witness. "Now then, you have been conducting private consulting over the past few years."

Carter admitted that was true.

"Who did you consult for, Doctor?"

"Your Honor, Coles objected, "I don't see the relevancy."

Reluctantly, Henderson overruled the objection.

"Would you answer my question, please," Brad requested.

"Well, various firms."

"To be more specific, Doctor, all, or nearly all, of your consulting has been for companies that have aggressively sought new wetland areas for construction sites. Is that not so?"

Carter looked uncomfortable. His eyes darted to the jurors to Coles and Cobb. "Yes."

"And how many of these companies received reports that discouraged building on the selected sites?"

"Objection, Your Honor," Cobb cried. "I don't see what bearing this has on the Riberia Street project."

"Sustained," Henderson intoned.

"I'll rephrase," Brad said. "Did any of your environmental assessment statements provide evidence that harm would come to the sites in question, all which were wetland sites?"

"Again I object," Cobb called out. "The question is the same."

"I must sustain," Henderson intoned.

"I'm sorry," Brad said, fighting to control his temper.

"Can you tell us what your reports showed?"

Carters licked his lips. "Most indicated little or no harm to the environment."

"Most? How many?"

"Well, I think they all showed no harm."

"You claimed in your assessments that construction would not harm them?"

"Yes," Carter replied.

"In summary, then, isn't it true that you consulted primarily for firms wanting to use wetlands sites for construction and in not one instance did you issue a negative report?"

Carter nodded.

"Please speak up, Doctor."

"That is correct."

"And you were paid well for your services?"

"Your Honor, I object," Coles shouted.

"Sustained. Disregard the question," Henderson told the jury.

"No further questions," Brad said.

When Carter returned to his seat, Henderson asked if the plaintiff attorneys wished to call other witnesses. They did, a Dr. Jerome Scanlon from the University of Miami. Scanlon, a short, chubby man with a checkered vest under his sport coat, smiled condescendingly as he took the witness stand.

"Doctor," Coles began after Scanlon was sworn in, "your expertise is in marine biology. Is that right?"

"Yes sir."

"Do you agree that putting up condos on the Riberia Street site will create devastation to the wetlands?"

Scanlon smiled knowingly. "No. The wetland in question is an isolated swamp suitable for not much of anything. Well, let me correct myself; that's not entirely true. It is suitable for construction. Especially, with so much land being taken up for housing developments, parking lots, and shopping malls. We need all the available land we can get."

"I have one further question, Dr. Scanlon. You heard what Mr. Larson said about the wetlands being connected in some strange way."

Brad objected. "The phrasing isn't entirely correct, sir."

"We'll let it stand," Henderson replied. "Continue, counselor," he said, looking at Coles."

"We're waiting for your opinion, Doctor," Codes prompted.

With a grin playing around his mouth, Scanlon replied that the entire concept was more like science fiction than true science.

"Are you saying all this interconnectedness is a matter of some scientists' imagination?"

"Well, I hesitate to put it that way, but as a scientist, myself, I find it pretty far-fetched."

Coles ginned and rocked back on his heels. "Well, thank you very much, Doctor." He glanced up at Henderson. "No further questions."

"Councilor?" Henderson said, directing his eyes upon Brad Longstreet and Shawna Gregory.

Shawna got up and walked over to the witness stand. She glanced down at a paper in her hand.

"Tell me, Doctor, what is your assessment of the Riberia Street wetlands?"

Scanlon frowned. "I've already stated that."

"I assume you have surveyed the area."

"Well, yes."

"You thoroughly checked it over?"

"Thoroughly enough."

"You visited the sight, of course.?"

"Of course."

"So you are familiar with the species that live there, the plants, trees?"

"I should think so."

"You walked through the area and examined the various life that inhabits the site?"

"I think I answered that."

"From what vantage point did you inspect the area?" Shawna pressed.

Scanlon hesitated. "From the road."

"The road into the wetland site?"

"Yes."

"You did not go any further?"

"There was no need to."

"Why is that?"

"What I saw was sufficient."

"Sufficient for what, Doctor? Sufficient to testify as to what results construction would have?"

"Yes."

"How could you do that? How could you just look at a wetland from a distance and decide what impact construction would have?"

"Experience."

"What species of crab live there?"

"Species? Well, I imagine..."

"No, Doctor, please be specific. Did you see any insects or spiders? Did you notice any terrapin, clam worms, shore shrimp, or mud snails?"

"That would have taken a lot of time."

"If you don't know what species live there, Doctor, how can you say that construction will or will not harm these?"

"Well, I don't see..."

"You get paid for testifying, don't you, Doctor?"

Coles rose to his feet. "Now judge, I must object to this line of questioning on the grounds it is irrelevant."

"Sustained."

"You testified," Shawna said, "that Sheldrake's hypothesis and theories that arise from quantum physics are, I think you said, far-fetched."

"That's correct."

"Have you studied quantum physics, Sir?"

"I've read a little about it."

"And you are familiar with Sheldrake's hypothesis?"

"Familiar, yes."

"Which of Sheldrake's' books have your read, Doctor?"

Scanlon squirmed in his chair. "I don't recall the titles."

"Was one of them, perhaps, The Morphogenetic of Species?"

"Yes, I believe that was the title."

"Are your sure, Doctor?"

"Yes, I'm sure," Scanlon replied testily.

"That was not one of Sheldrake's book, Doctor. In fact, I'm not sure it is the title of any book. I made it up."

Sweat beaded on Scanlon's forehead.

"It's hard to recall everything I've read."

"Tell me, what does Sheldrake mean by morphogenic fields?"

Scanlon looked about desperately. "I'm not sure he means fields in the scientific sense."

"What does he mean, do you think?"

"Well, morphogenic fields would be a complex event."

"Can you explain?"

It was obvious Scanlon was now over his head and not all that familiar with Sheldrake's ideas.

"I don't wish to take the court's time for a lengthy discussion," Scanlon bleated.

"The fact is," Shawna cut in, "you know very little about Sheldrake and his morphogenesis."

"I object, Your Honor," Coles shouted. "She's badgering the witness."

"Sustained."

"I have no further questions, Your Honor," Shawna said.

Henderson asked if the defense had any other witnesses to call.

Coles said they did not.

"In that case," Henderson said with a wrap of his gavel, "court adjourned until one o'clock tomorrow afternoon. At that time the court will hear the closing arguments."

CHAPTER TWENTY-SEVEN

"Is there any chance you could be in danger?"

Shawna recalled the looks on the Gears faces in court. Cody had ridden with her to her place. They were seated at the kitchen table at Shawna's discussing the day's trial. It was dark now. The wind blew the rain hard against the house. Lightning lit up the outside and thunder boomed and rumbled in the darkness.

"Danger?" Cody asked.

"From the Gears."

"They aren't stupid."

"What about the beating? That wasn't very smart." She paused. "I don't want anything to happen to you, Cody."

They ate a quick supper of hamburgers and fries, because Cody had to get back to his boat to make sure everything was secure. He had followed Shawna home on his Harley after the trial.

Cody glanced toward a kitchen window. "I don't want to be out if this wind gets any stronger."

"It doesn't look very good, Cody. Why don't you let me drive you?"

He got up and kissed her cheek softly. "Thanks. I'll be okay. No sense in you being out in this, too."

She stood in the rain-swept doorway while he cranked up the Harley, then ducked back in, waving to him as he went down the driveway.

Numerous limbs now littered the road. Despite his waterproof slicker, the torrential downpour soaked him by the time he swung into the marina.

On board, he changed into a bathrobe and mixed an Old Fashioned. Then he lay on his bunk as he listened to the weather report.

A hurricane reconnaissance aircraft reported Eddie's winds had increased to 70 mph. If they reached 75, the weather bureau would upgrade the storm to a hurricane. Forecasters still expected Eddie to remain offshore as it followed a northerly course. Because such storms can be very unpredictable, forecasters advised persons living near the coast to take precautions against wind damage and flooding.

"A little late," Matheson mused, as he listened to the wind screaming through the rigging, rain pelting the Columbia's cabin, and thunder rumbling across the night.

He reached for a book he'd started and read only two pages when the sound of his name startled him. He wondered who would be looking for him out in this storm like this. The tone of the man's voice was not friendly

He rose up from his bunk and stuck his head into the companion way. "Who is it?" Matheson shouted. The wind-driven rain beat against his face, making it difficult for him to see who it was.

The blast from the shotgun barely missed him. As he recoiled, he fell back into the cabin.

"Jesus Christ," he shouted.

He threw off his bathrobe and pulled on his pants, but not before the row of cabin ports exploded inward, sending shards of Plexiglas and fiberglass throughout the cabin. Some cut his face.

"For Christ's sake," he cried, "what the hell's going on?" He lay on his stomach with his hands over his head.

The reply was by another volley. As he dared glance up, he saw two pairs of legs through the shattered ports.

Another round of gunshot blasts. Buckshot splattered all around the cabin.

Damn, he didn't have anything to defend himself with. The only possible implements of possible defense were the fire extinguishes. But what in hell good would they be against the goddamn shotguns? Who were these guys?

"Hold your damn fire," he yelled. "Tell me what you want."

"You ain't long for this world, you fucker," one of the men shouted back.

Why in hell doesn't someone come? The sound of the shooting should certainly bring help.

"Tell me what you want, damnit," he shouted.

His reply triggered two more rounds fired simultaneously. The ports were now knocked out. The shots left gaping openings through which Matheson could see the wind-whipped pants of two men as the wind blew rain into the cabin.

Another sound.

A woman's voice.

"Oh, Christ, no!" Cody shouted.

"Professor," Jeanette called again. "What's going on?"

"Get back, get back!" Cody yelled at her. "For Christsake get below!"

He heard a single shotgun blast.

"No!" he screamed.

Cody leaped up.

"You goddamn sonsabitches," he roared.

Frantically, he seized one of the fire extinguishes and, without concern for his safety, he jumped into the companion way, into the rain, the wind, the vivid lightning, the bombardment of thunder—and sudden death.

With the double-barreled shotguns aimed directly at him, he fired the extinguisher point blank at the two men, He knew the force of the deadly pellets from the shotguns would slam him backwards, dead before he hit the cockpit floor.

But Lord, he pleaded, give me time to squeeze this off.

He knew the foam would be ineffective, but this was his last desperate effort.

The two shots came almost simultaneously. Cody stiffened as he pulled the extinguisher's trigger.

He felt nothing. Instead, he saw the two gunmen pirouetting in slow motion, covered with white foam. Their knees buckled, and in the flashes of lightning and dock lights, he caught their last moments, their final expressions of fear and disbelief.

Pazzio came at a run from the end of the dock, a pistol in each hand. In his black foul-weather gear, he appeared like the phantom of death.

Cody snapped his gaze back to the two men lying in the rain. But only for a moment. Frantically, he leaped onto the dock and jumped aboard Jeanette's boat. She was lying sprawled in the cockpit. Her eyes were open. Her expression registered the final horror of the moment before she died. Half her left shoulder had been torn away. Blood was all over the cockpit.

Before Matheson could touch her, Pazzio shouted, "Get her below." His eyes were ice. His face, streaked with rain, was grim, like the death he dispensed.

"What in hell's going on?" Cody cried.

Tears streamed down his face and the rain blinded him as they carried her down the companion way and laid her on a bunk.

Pazzio covered her with a blanket.

"Oh Christ," Cody moaned. "She came up on deck. I yelled at her to get below."

"I should have known," Pazzio said.

Cody stared at him.

Pazzio glanced around, grabbed a pencil, and paper and wrote quickly.

"Call this number. Tell Vito what happened." He went back up on dock and ran back through the storm to his boat.

Cody did not understand. He looked at the name and number. He thrust the paper into his pocket and ran down the dock after Pazzio.

"I'll get the lines," Cody shouted, instinctively knowing what Pazzio had in mind. As he untied the mooring lines, Pazzio started the boat's engines.

Cody was vaguely aware there were others now crowding onto the dock. A group had formed around the two gunmen, Others were milling about, bracing themselves against the storm.

The sound of sirens came down Riberia Street as Cody released the last mooring line and tossed it on deck. The forty-five foot ketch slid away from the dock. Pazzio at the helm. Cody watched it move into the river and disappear into the darkness without its running lights on.

"What the hell's happening?"

Mitch stood beside him, clawing at his crotch.

"I heard over my scanner there was some shooting. I find your boat's blasted to hell, two dead guys on my dock, someone telling me Jeanette's been shot. And where in hell is Pazzio going in this weather?"

"I'll explain later."

He followed Cody back to where a group had formed around the two men sprawled in the rain.

Mitch gazed at the gaping holes in Matheson's boat and told a couple men to get some tarps to put over the cabin.

Suddenly, a voice barked with authority. "All right, let's clear the area."

Cody saw Riker approaching with two men.

At the same time an ambulance, lights flashing and siren cutting the air, swung into the marina. Cody ignored Riker as he ran toward the two men who climbed out.

Cody pointed at Jeanette's boat. "She's below. Get her to a funeral home."

"Who?"

"Never mind, goddamnit."

"What about those two guys," one of the medics asked.

"Damnit, I said get the woman."

"Hey, you just wait," Riker shouted at Cody, "Who the hell are you to give orders?"

At that moment the sheriff's car careened into the marina, throwing mud all over Riker's patrol car. Randy Banks jumped out along with a deputy, who struggled to hold an umbrella over his boss's head. The wind turned it inside out. Banks called to Riker, "What's all this about, Clarence?" Then seeing the two bodies, he ordered, "No one touch anything, hear?"

The two officers conferred briefly, pausing to watch the medics put Jeanette's body into the ambulance.

"Who's that?" Riker asked.

The sheriff shook his head as Riker shouted at the medics. "Listen you guys, I want a report?"

Sheriff Banks pointed to the two men lying on the dock. "We got a couple stiffs. Looks like someone shot 'em."

"Well, we better get to the bottom of this," Banks said. He looked at the people standing around. "Who knows what happened?"

No one answered.

Another ambulance pulled up. "Hey," Banks shouted, "you guys take care of these two."

Not wanting anything to do with either Banks or Riker, Cody inched away. Neither saw him leave and hurry toward the marina's outside phone,

Before he finished dialing, Shawna swung into the marina's entrance.

She was out of her car instantly. "You all right? I heard over the scanner there was a shooting." Her face was white in the wind-swept rain..

"I'll explain later. Let's get the hell out of here."

Without questioning, she got back in her car and, as Cody jumped in beside her, she drove fast out through the gate and down Riberia Street.

They drove through the storm, down King Street and across the Bridge of Lions. Brilliant lightning illuminated the low dark clouds. Thunder crashed and rumbled across the violent night. Trees thrashed in the wind as branches flew through the darkness, littering the streets like corpses snatched away by the invisible hands of the storm. The car rocked in the violent gusts of wind.

"She's dead," Cody blurted finally.

"Shawna glanced across at him. "Who?"

He shook his head. Tears ran down his face. "The sonsofbitches killed Jeanette. For no goddamn reason. No goddamn reason at all. Christ all mighty." He stared straight ahead through the rain-blurred windshield. Suddenly, he smashed his fist on the car's dash in a frenzied outburst and buried his face in his hands.

CHAPTER TWENTY-EIGHT

Shawna lit candles, because the storm had knocked out the electricity. Then she mixed Cody a stiff drink and one for herself. He stood by the kitchen window and listened to the storm. Tree limbs littered the yard.

He had just finished telling Shawna everything that happened.

Finally, as Shawna held a candle, Cody dialed the number Pazzio had given him. The man who answered wanted to know who was calling. "It's about Jeanette."

Silence.

"What about her?"

"Look, put Mr. Severino on."

"You gotta be more explicit, Pal."

"Damnit, she's been shot."

A longer silence.

Then, "Who is this?" It was a different voice, more authoritative.

"Mr. Severino?"

"Talking."

"Tony Pazzio gave me your number."

"What about Jeanette?"

Cody hesitated, struggling for the words. There was no way he could soften the news. "She was shot." He heard a sharp intake of breath.

Cody explained what had happened.

"Where is she?"

Cody told him in a funeral home. He'd get the name.

Silence.

"How do I reach you?" Vito Severino asked.

Cody gave him Mitch and Shawna's numbers.

"I'm so sorry," Shawna said when Cody hung up.

"It's so goddamn senseless," Cody said angrily, as he downed his drink. Shawna mixed another.

"Could the Gears be behind this?" She asked

"They have to be."

"If so, they must be scared. They may have had doubts about the outcome of the trial."

"How the hell we going to prove they're involved. Like everything else they've done, we have no way to pin it on them."

Cody took a big swallow of his bourbon and water.

"One way or the other," he declared, "I'll get them for this."

They talked until after midnight then turned in. Cody slept fitfully as his mind reeled with all that had happened.

By morning, he got up to find Shawna fixing breakfast.

Although it was still raining, the worst of the storm had moved on, leaving in its wake downed trees, flooding, and broken windows. With breakfast over, Shawna drove Cody back to the marina. She had to maneuver around tree branches and downed power lines. St. Augustine was all but closed down. Work crews had started the clean-up. Some roads were still closed.

"Better get on home before they close any more roads," Cody told Shawna, when she pulled up at the marina entrance. "I'll ride the Harley out to your place a little later. As she left and he walked down through the marina he saw that the area had sustained little damage, except for one yacht that had been blown off its stands.

Cody retrieved from the Columbia a batch of notes he'd taken for articles. The men had done a good job covering the boat with a tarp. His notes were dry.

He then stopped by the office to see Mitch.

"It's a damned shame," Mitch said, "We know damn well who was behind the attempt on your life and Jeanette's murder.""

"Something will break. The Gears can't get away with what they've been doing forever."

"One of those guys is still alive," Mitch said. "He's in Flagler hospital. Maybe they can get him to talk."

" He's Lucky Pazzio didn't kill him."

"By-the-way," Mitch asked, "Did Pazzio tell you anything."

Cody didn't see any point relaying Pazzio's role in all this. "He took off right after the shooting."

"Why?. What he did isn't a crime. I doubt if he'd be charged. Certainly not with murder. He was protecting Jeanette, and probably you, too."

"That's right, but maybe he had his own reasons."

The Harley kicked over with the first try, and Cody swung down Riberia street and out to Shawna's.

As he drove down the roads now partially cleared of the debris, he found Jeanette intruding upon his thoughts.

The following morning, when he and Shawna had finished breakfast, the phone rang. It was Mitch.

"A guy here to see you."

"Who?"

"I didn't see who it was, but a guy came in and says his boss wants to see you. His name is Severino."

"Did you say, Severino?" Cody asked. The man must have flown in right after I talked with him, Cody thought. "Tell him I'll be right there."

As Cody swung into the marina and put the Harley on its kickstand, he saw a black limousine with tinted windows parked in front of the office. Two men in dark suits stood next to it.

In the office, Mitch nodded at the car. "That's him."

Vito Severino sat alone dressed in a dark suit in the back seat of the limo. He had graying hair and a black mustache. He looked haggard. As though he'd had little sleep. He gestured for Cody to get in. As he removed his dark glasses. Cody noted his deep blue eyes; like Jeanette's.

Severino stared at him for a few moments. "I want her sent home for the burial. They will see to it." He nodded toward the men outside the car,

"One still lives," he said.

The local paper had revealed that the surviving gunman had implicated the Gears. According to the story the district attorney, Mike Reese, had issued an indictment against the Gears, arguing unsuccessfully before Judge Henderson against bail.

"You have no boat now," Severino said.

"I can get it repaired," Cody said.

"You can have Jeanette's."

The offer surprised him. "I can't accept it. The boat is, yours, now."

Severino raised his hand. "You keep it or sell it. I want nothing of it. I'll have the papers drawn up."

Cody thought of Jeanette's plan to sail to the islands.

"You are also short of funds."

"I'm doing okay," Cody replied, wondering how much Severino knew about him. "I've sold a couple articles and am working on more. They may not pay a hell of a lot, but they keep me in enough money to get by."

"This will help." He withdrew a roll of bills from his jacket pocket. When Cody refused, it Severino thrust the money into his hand. Later, when Cody counted it, Severino had give him ten thousand dollars.

"I'll pay you back."

"You will not."

Vito motioned with his hand and the door opened. "I wish you well. She wrote about you,"

With that, their meeting ended.

That was the last Cody saw of Jeanette's father. It was not, however, Severino's last act before leaving St. Augustine.

Later, as he'd promised Vito, Cody packed Jeanette's personal belongings and mailed them to the address Vito had given him. It had been difficult going through her things, as though he were prying into her life. But while he removed her belongings, Cody felt that somehow she still remained. For this reason, he could not immediately move aboard her boat, so he accepted Shawna's offer to stay at her place.

In the wake of Jeanette's death, Cody had difficulty getting his thoughts and emotions back in order. Again and again he was brought back to that horrible moment when she came on deck as the gunmen were blasting his boat. And there was Water Rat's death, Pazzio's role in shooting the two gunmen, and the appearance of Jeanette's father.

Cody had no interest in anything for the moment, He began to drink more than usual. Shawna worried about him, but said nothing, hoping he would soon snap out of it.

While staying with Shawna, he rewrote the latest article. Instead of the old manual typewriter he had borrowed from Mitch, he now used a lap-top bought with the money from Severino.

As he put down his thoughts, he found that writing became a catharsis. It drew him out of his dark mood. Instead of trying to run from Jeanette's death, he accepted it. He saw his writing as a tribute to her. How often had she told him he should write a book? But this wasn't a book, just a series of articles. He didn't think that would bother her.

Initially, he worked at Shawna's. But as he reworked his articles, he moved to Jeanette's boat—he still couldn't think of it as his—where he found it easier to write, being closer to his subject matter, the wetlands.

When he paused now and then to look out across the wetlands and watch the variety of life flourishing around him, he found a new contentment, a growing feeling of fulfillment. He experienced something else, it was like a déjà vu. He felt Jeanette watching him, standing by the mast or sitting in the cockpit. Sometimes alarmed by the strength of his feelings, he would glance up. But each time he saw only the forest of masts and the wetlands bordering the Sebastian River. He spoke to her, thinking it sounded stupid, and hoped no one heard him. Then he read aloud what he'd written, as though asking her if she liked it.

Sometimes he wondered if he was losing his mind. He'd seldom thought seriously about ghosts or spirits or life after death. But now he wondered if the soul might live on, and the departed could somehow communicate with the living. Somewhere he'd read that the spirit remained for a while after death, but then moved on.

When her presence became overpowering, he stopped writing and studied the wetlands and watched the hermit crabs scuttling in the mud, and the storks circling above the cord grass dancing to the sea breeze. It sounded like Jeanette whispering to him.

"You think about her a lot, don't you, Cody?" Shawna remarked one evening after supper. They were on the back patio. The last rays of the sun thrust golden lances eastward across the sky.

Cody looked at her a moment. "Yes. It's almost like she's helping me write."

"That's why you don't work here?"

"I need to be near what I'm writing about. But I can't let myself become possessed. She's gone, I don't want to think I'm subconsciously trying to bring her back."

Shawna reached over and took his hand. Her smile was warm and loving.

The next day she told him the Gears were gone.

"Gone?" he asked.

They had simply disappeared, she said. According to what she heard at the office, a neighbor reported he hadn't seen either of them and that their car was in the driveway. Cordova and Phillips reported neither Ed nor his wife had been to the office for three days. They'd never stayed away that long.

"You think they skipped town?" Cody asked.

"We don't know."

Riker and Banks reported they thought the Gears had left to avoid prosecution. Both the sheriff and police chief put out APBs. Nothing turned up.

"They won't find them," Cody remarked.

"You know something?"

"I have a feeling."

"Severino?"

He nodded. "Look at it this way, no one leaves without taking something, a few clothes, money, toilet articles."

The disappearance of the Gears added another twist to the futile attempts by Banks and Riker to find who shot the gunmen.

When the Gear's house was searched, nothing was missing. It was like they'd simple gone to the grocery store. Their empty suitcases were in a closest and no money had been drawn from their banking accounts.

Neither Banks nor Riker wanted to talk about it. Nor did they want to comment on the death of the hospitalized gunman. The reason was obvious. According to the Record two men identified themselves as FBI agents to the uniformed guard outside the man's hospital room. When they emerged, they thanked the guard, shook hands, and left. Later, when a nurse checked on the patient, he was dead, strangled.

While the death of the second gunman eliminated the need for one trial, the Gears still had to stand trial; that is. if they could be found. Twice Henderson postponed the date, expecting the police would find them. When they didn't, he postponed it indefinitely, and took off on his vacation.

As the rumors flew about the disappearance of the Gears, those who had been involved with the Gear's condo project disavowed any connection with it. Through his press secretary, Victor Blanchard issued a statement that, while he originally thought the project good for St. Augustine, he had serious questions now. By a reconstructing the truth and employing false imputations, he appeared innocent of any complicity in alleged illegal activities of the Gears.

Mayor Douglas Orlando. was the second person to divest himself of any connection with the project. According to the Record he resigned because of poor health. Within a week he left town.

Two others who had left ahead of him: Juan Cordova and Sidney Phillips.

City Manager Brad McCarver declared the city needed a younger man, "fresh blood," as he put it, and stepped down to pursue other activities. Commissioner Wally Furgerson, faded from the political scene like a puff of evaporating smoke.

But city attorney, Kent Daniel, saw the resignations as an opportunity to further his own political interests. He immediately put in his bid to run for mayor. He hoped one day to become the district attorney. Fortunately for his career, he had covered his tracks sufficiently to remove evidence of any involvement with the condo project.

CHAPTER TWENTY-NINE

Cody sat under the canopy in the cockpit of what had been Jeanette's boat. The sun was setting. As he finished the article and put aside his lap-top, he looked out across the wetlands and wondered what would become of them, now that the condo development project was on indefinite hold. After all that had happened, the city fathers showed no interest in the future of the project.

While working, he had been acutely aware of the ineffable presence of Jeanette. Now that the piece was finished, he felt a sudden emptiness, as though she were no longer present.

The mood passed, as he stacked the sheets and closed his laptop. He recalled the phone call from Gannet this morning. It came as something of a surprise, since he hadn't heard anything from him.

"I have some news I think you'll like," the congressman said.

"Try me," Cody replied.

After they'd hung up and he walked back to the boat, Cody wanted to call Shawna immediately and tell her what the congressman had said. However, he didn't want to bother her at her office, so would wait until she stopped by after work. There was something else he wanted to tell her. He had received a letter from a book publisher. The editor was interested in his idea and wanted more information. Although the letter implied nothing definite, it was, nevertheless, encouraging. He'd sent the proposal for a book to several publishers, but despaired that he would get any bites. He had said nothing about it to Shawna.

On the face of what the congressman relayed, Cody felt Gannet's plan had great possibilities. Aboard Jeanette's boat, he mixed a drink and sat in the cockpit enjoying the late afternoon. When he'd finished his drink, he reviewed an article about the wetlands he'd been working on. Finally, he checked his watch. Shawna would be by soon.

As movement caught Cody's eye, he looked up to see Shawna coming along the dock.

"Come aboard," Cody said, jumping down and taking her arm. "I got some news. But first, how about a drink."

"That sounds good."

"We're celebrating."

"Oh? What?" She asked, adjusting a cockpit pillow to make herself comfortable.

"In a minute," he replied, as he went below to make their drinks.

When he returned, he said he'd received a call from Gannet.

"Oh? When?"

"This morning."

"He said he'd given considerable thought to our conversation at Scarlet O'Hara's. He agreed that more legislature isn't the answer to preventing a continued onslaught of our environment, particularly the wetlands."

Cody sipped his drink. "He apparently went along with my idea. And apparently he has the backing from other congressman. He thinks it's likely he can get approval for some $300 million for the various approaches I outlined during our meeting with him."

"Wonderful!" Shawna cried as she jumped up and threw her arms around his neck, causing him to spill some of his drink.

When she'd sat back down, Cody added, "There's more."

She regarded him quizzically.

"The Riveria Street wetlands may be designated as one of several to be purchased with state and matching municipal funds. That's still in the works, Gannet said."

"That's great," she cried.

"There's more."

She scowled at him. "Let's hear it, Cody,"

He took another drink before answering. "I'd say it's an excellent reason to celebrate.

I got a letter from an editor at Crown Publishers. It was in response to a proposal I sent out. I'd also included a copy of an article I did on the wetlands. He's interested."

"Cody, you can be exasperating. What proposal are you talking about?"

"That I write a book, using some of the idea in the articles I've written."

"You mean a book about the wetlands?"

"Well, not exactly."

"What then, exactly?"

He finished his drink, took her's and went below.

When he returned, he said, "I've been doing some thinking."

"What about?"

"Us." He paused to sip his drink. "And Jeanette's desire to photograph the islands."

"I know," Shawna said softly. "You were very fond of her."

"She would have left for the islands, except for me."

"She was a wonderful person, Cody. But you must not blame yourself for what happened."

"You're right, I know, but —"

"Cody, you can be exasperating. What about us?"

"I was afraid you wouldn't be interested."

"Why?"

He shrugged. "I thought at first we might buy a place overlooking the wetlands. I want to wake up in the morning and see them. I want to study them under the changing light and the changing seasons."

"You're thinking of a place for the two of us?"

"That was at first."

"Then what?"

When he didn't answer immediately, she said, "Cody, will you get to the point?"

"I'm not sure how you'd react."

"There's only one way to find out."

"Okay," he said, taking a couple swallows of his drink. "I want to do what Jeanette had planned. But add something."

"You want to take pictures of the islands?"

"She wanted to sail the Caribbean and capture it on film. I would like to fulfill that dream, sort of for her, only I'd write a text to go with the photos."

"Like one of those picture travel books?"

"No. I'd focus on the environment I'd research the relationship between the people throughout the islands and their land. What they've done with it. I'd look for some symbiotic relationship, take photographs, and write about it."

When he finished, she thought a moment.

"How long would you be gone."

He hesitated. "Not me; we."

She stared at him. "You're saying you want me to go with you as you sail the Caribbean?"

"That's it."

For a moment she said nothing. Then she laid her hand on his. "Cody Matheson, is this a proposal?"

"Yes."

"And you're saying we'd honeymoon in the Caribbean?"

"Why not?"

She smiled. "I wouldn't like the idea of being separated from you."

"Just think," he said, grinning, "we'd photograph and write during the day and make love during the tropical nights."

"Oh?" she said, smiling. "Do you think you're up to that, the love-making, I mean?"

"I'd like to try." Then he added. "I broached the idea to the publisher who was interested in my book. He liked it, but suggested we talk about that after the book is finished. While photographing the islands, I'd work on the book.

She smiled. "In the meantime, we have a lot of planning to do. I have to close up the house, Tell, Mike I'm leaving. Settle our finances. That's just for starters."

"Then you'll do it?"

"Did you think I wouldn't?"

He finished his drink.

"I used some of the money Severino gave me to have the Columbia repaired. I may have a buyer. We'd go in this boat. It's in better shape. Although there is work to be done on it."

"You've given me quite a bit of a surprise and a lot to think about," she remarked finishing her drink. "In the meantime, I'm hungry."

"Let's go to a nice restaurant," he suggested,

"And afterwards, Cody Matheson?

He raised his glass and smiled. They sat looking at each other.

After he'd taken their glasses below, he helped her off the boat, and they walked up through the marina.

"You know something?" She said, looking up at him.

"I'm not sure."

"I love you, Cody."

A light was on in the marina office. Inside they found Mitch doing some late paperwork. He motioned to the bulletin board.

"Got some new mail," he said.

It was covered with notes, cards, and letters. A note from Rocky and Millie said they had reached Trinidad. Gus Johnston and his wife had written from San Diego saying they were gearing up for their sail to Hawaii. Marian Talbert was in the Barbados. She hadn't encountered any hurricanes and was having a wonderful time. Jules Rodreques was having engine trouble in Charleston, SC. Jugs and her husband Jake wrote from Cozumel off the coast of the Yucatan. They were doing fine.

Cody turned to Mitch. "Just about everyone we knew has left."

"They'll be back."

"I'm sure they will. But we may miss them."

Mitch looked at him questioningly.

"What Cody is saying," Shawna put in, "is he more or less proposed to me."

"Well," Mitch exclaimed as he jumped up and grabbed Cody's hand. "Congratulations."

"Which means we'll be leaving, too. I'll let Cody tell you."

Cody explained about the book publisher. He also told him about Gannet's plan."

"Say, you old…" he was about to say fart, but in deference to Shawna he let silence fill the gap. "I'm glad for you, Cody. But you won't stay away forever. You got too many ties here in St. Augustine."

"We got a lot to do before we leave," Cody remarked. "But first, we got to celebrate and get something to eat."

Mitch came from behind the counter and hugged Shawna. "I'm glad for you both."

Outside, Cody glanced at the door to the men's room, remembering when Jeanette scrubbed the paint out of his hair.

There were many memories, but many unfamiliar boats now in the marina. A lot of new people. Strangers.

He paused a moment and recalled Chief Seattle's words: *This we know: the earth does not belong to man, man belongs to the earth. All things are connected like the blood that unites us all. Man did not weave the web of life, he is merely a strand in it….*

"Let's go." he said, taking Shawna's arm as they walked to her car.

He glanced over at the wetlands where the Gears had planned their ill-fated condos. In the dusk he watched an osprey settle into its nest with a trophy of fish.

Before opening the car door, he listened to the sibilance of the incoming tide—the wetland's music, accompanied by the wind sighing through the Spartina.